Breaking the Availability Barrier
Survivable Systems for Enterprise Computing

Breaking the Availability Barrier
Survivable Systems for Enterprise Computing

Dr. Bill Highleyman
Paul J. Holenstein
Dr. Bruce Holenstein

© 2004 by Dr. Bill Highleyman, Paul J. Holenstein, and Dr. Bruce Holenstein.
All rights reserved.

No part of this book may be reproduced, stored in a retrieval system, or transmitted by any means, electronic, mechanical, photocopying, recording, or otherwise, without written permission from the authors.

ISBN: 1-4107-9231-5 (e-book)
ISBN: 1-4107-9232-3 (Paperback)
ISBN: 1-4107-9233-1 (Dust Jacket)

Library of Congress Control Number: 2003099708

This book is printed on acid free paper.

Printed in the United States of America
Bloomington, IN

All products mentioned in this book are trademarks of their respective owners. The information in this book is provided on an as-is informational basis. The authors, owners, and publisher disclaim liability for any errors or omissions. The reader accepts all risks associated with the use of the contents of this book.

1stBooks – rev. 12/18/03

Dedication

This book is dedicated to our spouses,
Janice, Karen, and Denise,
for their enduring patience and support.

Breaking the Availability Barrier

Contents

Forward .... **xvii**
PART 1 – BREAKING THE FOUR 9S BARRIER xviii
PART 2 – ADVANCED TOPICS xxii
APPENDICES xxiii
AUTHORS' NOTES xxiv
ACKNOWLEDGEMENTS xxv
ABOUT THE AUTHORS xxv

Part 1 - Breaking the Four 9s Barrier 1

Chapter 1 - The 9s Game 3
WHAT IS RELIABILITY? 3
SOME CAVEATS 6
9S – THE MEASURE OF AVAILABILITY 6
TODAY'S SYSTEMS 8
SIMPLE SYSTEMS 8
   Non-Redundant System 8
   Redundant System 9
DOUBLE YOUR 9S 10
THE REAL FAULT-TOLERANT WORLD 10
RANDOMLY DISTRIBUTED PROCESS PAIRS 11
PROCESS/PROCESSOR PAIRING 13
AVAILABILITY IN GENERAL 14
MORE SPARING 15
HOW MANY FAILURE MODES? 16
THE IMPACT OF REPAIR TIME 17
SOME HELPFUL CHARTS 19
ANSWERS 21

Breaking the Availability Barrier

SUMMARY 22
HOW FAR SHOULD WE GO? 23
A CASE STUDY 24
WHAT'S NEXT? 25

Chapter 2 - System Splitting 27

THE AVAILABILITY RELATION 27
FULL REPLICATION 29
SYSTEM SPLITTING 31
   Simple Splitting 31
   Multiple Splitting 34
   Impact on Mean Time Before Failure 36
   Elimination of Planned Down Time 37
REPLICATION OF DATA 37
MUST WE REPLICATE THE DATABASE? 39
SUMMARY 42
SYSTEM SPLITTING AND PROCESS PAIRING 43
WHAT'S NEXT? 44

Chapter 3 - Asynchronous Replication 45

USES FOR DATA REPLICATION 46
   Disaster Tolerance 46
   Increased Availability 46
   Localization 46
   System Maintenance 47
   Enterprise Application Integration (EAI) 47
   Operational Data Store (ODS) 47
   Data Warehousing 47
   Zero Down Time System Migration 48
TYPES OF DATA REPLICATION 48
   Directionality 49
   Threading 51
   Queuing 52
   Target Updating 53
   Synchronism 54
THE BASIC DATA REPLICATION ENGINE 56
ADVANTAGES OF ASYNCHRONOUS REPLICATION ... 59

Breaking the Availability Barrier

No Performance Penalty 60
Non-Invasive 60
Heterogeneous 60
Data Manipulation 60
Data Integrity 61
Highly Reliable 61
Highly Available 61
Highly Scalable 61
Highly Secure 62
ASYNCHRONOUS REPLICATION ISSUES 62
General Issues 62
   Data Loss 62
   Database Corruption 63
Ping-Ponging 64
Data Collisions 66
COLLISION AVOIDANCE 67
Partitioned Database 67
Synchronous Replication 68
COLLISION DETECTION 68
Before-Image Comparison 69
Versioning 69
COLLISION RESOLUTION 69
Generic Algorithms 69
Data Field Specific Algorithms 70
Relative Replication 70
Fuzzy Replication 71
Manual Resolution 72
FAILURES AND RECOVERY 72
Failover 72
   Source Node Failure 72
   Target Node Failure 73
   Network Failure 73
Restoration 74
   Duplicate Transactions 75
   Data Collisions 75
WHAT'S NEXT? 76

Breaking the Availability Barrier

Chapter 4 - Synchronous Replication 77
REPLICATING SYSTEMS 77
SPLITTING SYSTEMS 78
DATA COLLISIONS 80
SYNCHRONOUS REPLICATION 81
   Dual Writes 82
   Coordinated Commits 84
APPLICATION LATENCY 86
   Dual Writes 87
   Coordinated Commits 88
SYNCHRONOUS REPLICATION EFFICIENCY 89
SCALABILITY AND OTHER ISSUES 93
   Scalability 93
      Multiple Database Copies *93*
      Communication Channel Efficiency *93*
   Transaction Profile 94
   Read Locks 95
   Other Algorithmic Optimizations 95
EXAMPLES 96
   Geographically Distributed Systems 96
   Collocated Systems 98
EFFICIENCY MODEL EXTENSIONS 98
   Dual Write Single Round Trip Operations 98
   Dual Write Serial Updates 99
   Plural Writes 101
DEADLOCKS 102
FAILURES AND RECOVERY 103
   Dual (Plural) Writes 103
   Coordinated Commits 103
      Failover *103*
      Restoration *104*
WHAT'S NEXT? 105

Chapter 5 - The Facts of Life 107
A REVIEW OF AVAILABILITY 108
WHY DO COMPUTERS STOP? 109
SOME DEFINITIONS 113

Breaking the Availability Barrier

TRIGGERED OUTAGES 114
THE IMPACT OF FAILOVER FAULTS 115
   A Better Value for Subsystem Availability 117
   Effect of Failover Faults on System Availability 118
   Effect of Failover Faults on Effective Subsystem Availability .. 118
   Effect of Failover Faults on System Splitting 119
THE GOLDEN RULE – REDUCE RECOVERY TIME 120
THE IMPORTANCE OF RESTORE TIME 122
WHAT'S NEXT? 124

Chapter 6 - RPO and RTO 125

REPLICATING THE SYSTEM 125
RTO AND RPO 126
REPLICATING THE APPLICATION DATA 128
   No Replication 129
      No Replication, Periodic Backup Only 130
      No Replication, Periodic Backup with Audit Trail 130
   Unidirectional Replication 133
      Unidirectional Replication – Cold Standby 133
      Unidirectional Replication – Warm or Hot Standby 133
   Bi-Directional Replication 133
   Active/Active Replication 134
      Partitioned Active/Active Replication 135
      Asynchronous Active/Active Replication 136
      Synchronous Active/Active Replication 136
RECOVERY TIME 137
   No Replication 137
      No Replication, Periodic Backup Only 137
      No Replication, Periodic Backup with Audit Trail 139
   Unidirectional Replication 139
      Unidirectional Replication – Cold Standby 139
      Unidirectional Replication – Warm or Hot Standby 140
   Bi-Directional Replication 140
      Hot Standby 140
      Partitioned Active/Active Replication 140
      Asynchronous Active/Active Replication 141
      Synchronous Active/Active Replication 141

Breaking the Availability Barrier

DATA LOSS 141
   No Replication 142
   Asynchronous Replication 142
   Synchronous Replication 142
RECOVERY STRATEGIES 143
   Unidirectional Replication 144
      Asynchronous Replication *144*
      Synchronous Replication *144*
   Active/Active Replication 145
      System Failure *145*
      Network Failure *145*
COMPARISON SUMMARY 146
MULTI-NODE APPLICATIONS 150
RECOVERY DECISION TIME 150
SUMMARY 151
WHAT'S NEXT 152

Chapter 7 - The Ultimate Architecture 155

AN AVAILABILITY REVIEW 156
THE STRAWMAN SYSTEM 158
SPLITTING INTO INDEPENDENT SYSTEMS 159
SYSTEM SPLITTING WITH DUAL DATABASES 162
DO WE NEED TO REPLICATE A MIRRORED DATABASE? .. 163
   Option 1: Split Mirrors 165
   Option 2: Network Storage 165
THE ULTIMATE ARCHITECTURE 167
PERFORMANCE IMPACT OF SYNCHRONOUS REPLICATION
.... 169
HERE COME LOCAL CLUSTERS 170
DATABASE REPLICATION – ENHANCEMENTS WANTED .. 171
CONCLUSION 172

Chapter 8 - The Rules of Availability 175

Breaking the Availability Barrier

Part 2 - Advanced Topics 179

Chapter 9 - Data Conflict Rates 181
SYNCHRONOUS VERSUS ASYNCHRONOUS REPLICATION 182
DEADLOCKS AND COLLISIONS 184
   Deadlocks 185
      Mutual Waits 185
      Lock Latency 186
      Intelligent Locking Protocols 186
   Collisions 186
THE MODEL 188
MODEL SUMMARY 190
   Synchronous Replication 191
      Mutual Waits 191
      Lock Latency 191
   Asynchronous Replication 192
      Transactions Sent Serially After Commit 192
      Transactions Broadcast After Commit 192
      Modifications Sent Serially After Application 192
      Modifications Broadcast After Application 192
      Modifications Broadcast Upon Receipt 192
MUTUAL WAIT DEADLOCKS 193
REPLICATION CONFLICTS 197
   Collisions Under Asynchronous Replication 197
      Transactions Sent Serially After Commit 199
      Transactions Broadcast After Commit 200
      Modifications Sent Serially After Application 200
      Modifications Broadcast After Application 201
      Modifications Broadcast Upon Receipt 201
   Deadlocks Under Synchronous Replication 201
      Mutual Waits 201
      Lock Latency 202
      Collisions, Waits, and Deadlocks 203
      Combined Effects 205
      Deadlock Resolution 207
PROCESSING NODES AND DATABASE NODES 210

Breaking the Availability Barrier

NOT ALL ACTIONS COLLIDE 212
FILE/TABLE HOT SPOTS 213
EXAMPLES 213
   Synchronous Replication 214
      Dual Writes *214*
      Coordinated Commits *214*
   Asynchronous Replication 215

Chapter 10 - Referential Integrity 217

BACKGROUND 217
   System Replication 217
   Data Replication 220
      Early Systems *221*
      Asynchronous Replication *222*
      Synchronous Replication *223*
      Physical Replication *224*
      Asynchronous Replication Issues *225*
   Transactions 226
   Simple Data Replication Model 229
   Natural Flow 232
   Referential Integrity 234
   Current Data Replication Architectures 238
      Single-Threaded Replication Engine *238*
      Single-Threaded Replication Engine With DOC *240*
      Multi-Threaded Replication Engine *241*
   Summary 242
MULTI-THREADING FOR PERFORMANCE 243
MULTI-THREADED EXTRACTOR 245
   RULES-BASED EXTRACTOR ASSIGNMENT 245
   Inter-Extractor Coordination 249
MULTI-THREADED COMMUNICATION CHANNEL 252
MULTI-THREADED APPLIER 254
   Suspend on Commit 254
   Appliers Coordinate Commits 256
   Using a DOC 257
EXCEEDING TRANSACTION COUNT LIMITS 258
   Multiple Appliers 259

Breaking the Availability Barrier

Partial Transactions 260
Adaptive Replication Engine 261
RESOLVING DEADLOCKS 262
Deadlocks With An Application 262
Deadlocks With Another Applier 262
Deadlocks With Another Transaction 263
SUMMARY 265

Chapter 11 - Failover Faults 271

FAILOVER FAULTS 272
REPAIR STRATEGIES 273
NO FAILOVER FAULTS 273
FAILOVER FAULT REQUIRING SUBSYSTEM REPAIR 275
FAILOVER FAULT REQUIRING SYSTEM RECOVERY 276
FAILURE MODEL SUMMARY 277
AN INTERPRETATION 278
Effective Subsystem Availability 278
System Availability Degradation 281
Splitting Systems 281
The Importance of System Recovery Time 285
MARKOV MODELING 286

Appendices 291

Appendix 1 - Availability Relationships 293

AVAILABILITY 294
SYSTEM SPLITTING 295
SYNCHRONOUS REPLICATION 295
FAILOVER FAULTS 297
DATA CONFLICT RATES 298

Appendix 2 - Availability Approximation Analysis 301

Appendix 3 - Failover Fault Models 305
GENERAL 305
NO FAILOVER FAULT 308
  Parallel Repair 308
   Sequential Repair 311
  Simultaneous Repair 314
FAILOVER FAULT REQUIRING SUBSYSTEM REPAIR 317
  Parallel Repair 317
  Sequential Repair 320
  Simultaneous Repair 323
FAILOVER FAULT REQUIRING SYSTEM RECOVERY 326
  Parallel Repair 326
  Sequential Repair 330
  Simultaneous Repair 333

Appendix 4 - Implementing a Data Replication Project 337
OVERVIEW 337
ANALYSIS OF NEEDS 338
EVALUATION OF OPTIONS 341
  Building a custom solution 341
  Buying a complete or semi-complete solution 342
  COTS Data Replication Products 342
  Distinguishing Criteria 343
RESOLUTION OF CONCERNS 344
PROJECT IMPLEMENTATION 345
COMPANY LISTING 346

References and Suggested Reading 349

Index 353

Forward

There is an old saying in business. You can optimize schedule, cost, and quality. Pick any two.

When it comes to configuring your data processing system, there is an equivalent saying. You can optimize performance, cost, and availability. Pick any two.

We typically configure our systems for performance and cost and let availability fall where it may. Even so, we are achieving impressive availabilities by industry standards. Typically, highly available systems such as HP NonStop Servers will fail only about once every five to ten years and will then be down for an average of about four hours. This means that our systems are up 99.99% of the time and are delivering four 9s of availability to our businesses.

When we talk about availability, we mean that all services upon which our businesses depend are available. "Available" means not only just working but also working at a performance level that makes our services useful. When sub-second response is expected, a multi-second response may be no better than no response at all.

Are four 9s good enough for you? You probably have faced what an outage of several hours means to your business. If you are running a Web service, this could mean irate customers, lost sales, and perhaps lost customers. If you are providing banking services, this may result in large fines by governmental authorities. A stock market trading system outage could make international headlines. The failure of a critical emergency service such as a "911" system in the U.S. could contribute to death by a cardiac arrest or a building burned to the ground.

How would you like to improve the availability of your systems so that the loss of any significant capacity is measured in terms of centuries rather than years – *at little or no additional cost*? As an added plus, your systems could tolerate major disasters such as floods, fires, earthquakes, and terrorism as well as environmental failures which may take out your power or air conditioning. Getting more availability for your system expenditures is what this book is all about.

This book is separated into two Parts. Part 1, *"Breaking the Four 9s Barrier,"* discusses the availability characteristics of today's systems and how to economically reconfigure these systems via distribution in order to dramatically improve availability. Part 2, *"Advanced Topics,"* presents a more detailed analysis of some of the considerations in distributing systems for improved availability.

The authors' intended audience includes IT executives who feel that they must reduce the down time of their systems, the system architects and senior developers who must build these systems or modify existing systems to achieve the required availability, and the operations staff who must then run these systems and recover from system faults. Many of the concepts presented are supported by mathematical arguments. However, for those who consider themselves mathematically challenged, we have tried to make clear the concepts being formulated so that the mathematics can, in fact, be glossed over.

Part 1 – Breaking the Four 9s Barrier

The availability of a system is directly related to the number of ways in which it may fail – its failure modes. We show in Chapter 1, *"The 9s Game,"* that the number of failure modes can be reduced significantly by paying attention to how we allocate critical processes to processors.

We show in Chapter 2, *"System Splitting,"* that failure modes can be further reduced by splitting a system into several smaller, independent nodes. This strategy not only dramatically improves availability but also, when a node outage does occur, guarantees that only the capacity provided by that node is lost. Furthermore, the chance of losing the capacity provided by two or more nodes is virtually never.

As an added advantage, splitting a system into several nodes allows you to do upgrades and maintenance a node at a time, virtually eliminating planned down time.[1]

However, nothing comes for free. If we split a system into several cooperating nodes, the system database must also be distributed across these nodes. Providing many duplicate copies of the database can be very expensive as often the cost of the database subsystem represents a majority of the system cost. Chapter 2 also shows how we can distribute a system across geographically dispersed nodes without suffering additional database costs.

Inherent in splitting a system into two or more independent nodes with duplicate copies of the database is that these databases must be kept in near synchronization. Chapter 3, *"Asynchronous Replication,"* shows how to do this by replicating data changes from the source node that is making the changes to the other database copies in the application network so that they can be updated with these same changes. We also explore in this chapter many advantages and several issues with asynchronous replication.

A particularly severe problem with asynchronous replication in some applications is data collisions that lead to database contamination. This occurs when two users at different locations simultaneously update the same data item on different database copies. The result is inconsistent data propagated across the network. Data collisions can happen with surprising frequency and often require manual intervention to resolve. The only general solution to this predicament is to avoid data collisions.

---

[1] It is interesting to note that one of the authors identified a need for a working peer-to-peer replicated system almost a decade ago. See Highleyman, W. H., *"Distributing OLTP Data Via Replication,"* Tandem Connection, Volume 16, Issue 2; April/May, 1995.

Dr. Bill Highleyman, Paul J. Holenstein, and Dr. Bruce Holenstein

Chapter 4, *"Synchronous Replication,"* describes how to avoid data collisions by ensuring that distributed copies of a database are kept in exact synchronism by the replication of updates across the network as atomic transactions. However, doing so can increase transaction response time because of communication network delays. The performance impact of different methods for synchronous replication is evaluated, and we show that the method of choice depends not only upon whether the system nodes are collocated or geographically dispersed but also upon the length of the transaction.

Distributed transactions (such as HP NonStop's network TMF) are generally appropriate for short transactions within a collocated distributed system. For geographically distributed systems or for large transactions, the use of asynchronous data replication with coordinated commits of transactions at each of the nodes is more efficient. In either event, the performance impact is more than overcome by the increased speed of new systems that are being introduced today.

Up until now, we have considered redundant systems whose outages are caused by multiple hardware failures. It turns out that more important factors are software faults and operator errors. Chapter 5, *"The Facts of Life,"* extends the availability concepts to include these sources of outages. Here we stress the importance of fast recovery from an outage and point out that this is not only a technical issue but also more importantly a serious business process issue.

As in every facet of life, there are trade-offs and compromises. When it comes to availability, two important considerations are the time that it will take for a system to recover from a failure and the amount of data that may be lost due to a failure. Every organization must grapple with their own tolerance to recovery time and data loss and set objectives for what can be tolerated. The corporate objective for recovery time is known as the Recovery Time Objective, or RTO. The corporate objective for acceptable data loss is called the Recovery Point Objective, or RPO. There are a variety of technologies today that allow one to replicate a system, and each has its own RTO and RPO characteristics. These various technologies are reviewed in Chapter 6, *"RTO and RPO."*

In Chapter 7, *"The Ultimate Architecture,"* we put all we have learned into configuring systems that can meet the following objectives:

- The frequency of losing more than a tolerable amount of capacity is measured in centuries.

- The system can be distributed geographically for disaster tolerance.

- The reconfigured system will incur little if any additional cost.

- Reconfiguring for availability is non-intrusive. It does not require application rewrites.

- By and large, the facilities for achieving these objectives are available today.

These are tough objectives indeed, and certainly there will be some cost and performance impacts. But are these impacts worth the significantly enhanced availability that can be achieved? Only you can answer that question.

Remember – a system that is down has zero performance and perhaps an incalculable cost.

Throughout these chapters in Part 1, a variety of rules that relate to very high availability systems are set forth. These rules are summarized in Chapter 8, *"The Availability Rules."* Perusing these rules is an excellent review and summary of the concepts presented in Part 1.

Dr. Bill Highleyman, Paul J. Holenstein, and Dr. Bruce Holenstein

Part 2 – Advanced Topics

Several concepts revolving around distributed systems are introduced in Part 1 of this book. But distributed systems pose many problems. Some of these are analyzed in more detail in the advanced topics of Part 2.

One of the problems in distributing an application and its database across multiple systems is that data collisions which contaminate the database may occur. A possible solution to this problem is to permit data collisions, then detect and resolve them as discussed in Chapter 3, "*Asynchronous Replication.*" As shown there, data collisions can in some instances be resolved automatically but in other cases must be resolved manually.

Another solution to this problem is to avoid data collisions by using synchronous replication, as shown in Chapter 4. However, this solution may present a performance challenge. In addition, if the applications in such an environment are not structured properly, deadlocks may occur when two or more applications attempt to lock copies of a data item, or attempt to lock the same set of data items in different orders.

Chapter 9, "*Data Conflict Rates,*" gives an analysis of data collision rates and deadlock rates that may help you decide whether you can tolerate data collisions in your application. It is shown that deadlocks are orders of magnitude less likely than data collisions.

Another potential problem in distributed databases is the violation of referential integrity rules if transactions are not replicated in natural flow order. The satisfaction of natural flow requires that transactions, and indeed the database modifications contained within the transactions, must be applied in the same order at the target databases as they were at the source database. Otherwise, for instance, a child entity might be added to the database before a parent existed; or a later update might be overwritten by an earlier update. Chapter 10,

"Referential Integrity," explains this problem and explores various data replication engine architectures and their ability to maintain referential integrity.

A major impact on the availability improvement that can be obtained by distributing an application is the fact that the automatic recovery from a fault may itself fail. This is called a *failover fault*. Failover faults are discussed in Chapter 5, *"The Facts of Life,"* where a cursory analysis of the impact of such faults on various aspects of system availability is presented. Chapter 11, *"Failover Faults,"* provides a more formal analysis of these impacts.

A fundamental requirement for increasing system availability by splitting a system into multiple independent nodes is a robust data replication engine that can support low-latency bi-directional replication, whether one is using a synchronous or an asynchronous approach. The authors have extensive experience in the field of data replication and can be contacted via the e-mail addresses given in their biographies for further information.

Appendices

As we progress through the book, we derive many useful relationships regarding the availability of distributed systems. These relationships are repeated and conveniently organized in Appendix 1 for ready reference.

Many of the relationships depend upon a fairly simple approximation given in Chapter 1, *"The 9s Game,"* to the dependency of system availability on subsystem availability, subsystem mean time before failure, subsystem mean time to repair, and the level of sparing. Such approximations are used to simplify the analysis and to give a clearer view of the concepts being presented. A formal analysis of these relationships is given in Appendix 2, *"Availability Approximation Analysis,"* where it is shown that the approximations not only are conservative but also are generally within 5% of the actual values over the range of parameters that are of interest to us.

Dr. Bill Highleyman, Paul J. Holenstein, and Dr. Bruce Holenstein

The Markov Chain models which support the various failover fault cases analyzed in Chapter 11 are detailed in Appendix 3, *"Failover Fault Models."*

Finally, Appendix 4 provides a methodology for implementing a data replication project and lists many of the data replication products that are available as of the writing of this book.

Authors' Notes

In many places throughout this book, reference is made to HP NonStop systems. NonStop systems were originally developed by Tandem Computers to provide very high availability. Tandem Computers was subsequently acquired by Compaq Computers, and Compaq was then acquired by HP. HP has changed the name of the Tandem systems to HP NonStop servers. The authors have considerable experience with these systems. However, concepts and recommendations presented in this book are extendable to all types of redundant systems, including HP Superdome, Windows Server clusters, UNIX clusters, and IBM Sysplex systems.

It should be noted that in the NonStop world, some important characteristics of fault-tolerant systems are summarized by the acronym RAS. RAS stands for reliability, availability, and scalability. In the RAS context, "reliability" is taken to mean "data integrity." For purposes of this book, the term *reliability* is used in the more general sense of system reliability. System reliability is associated with system availability.

Part 1 of this book is, in part, a compilation of a six-part series of papers published in The Connection, starting with the November/December, 2002, issue. The Connection is the official publication of ITUG, The International HP NonStop Users' Group. These papers have been modified slightly to fit the format of this book and include certain corrections. The originals may be found in the archives of The Connection at www.itug.org.

Each of the chapters in this book has been written to be self-standing at the risk of some repetition. Therefore, the reader is encouraged to pick and choose the topics of interest and to read only those chapters that apply. Adequate reference is made to other chapters to suggest further reading.

Acknowledgements

Breaking The Availability Barrier has benefited from reviews by many people. We gratefully acknowledge their guidance and especially thank Carl Niehaus, Wendy Bartlett, Alan Wood, Julie Scherer, and Gary Strickler for their in-depth evaluations and helpful criticism of various sections. We also thank Burt Liebowitz and John Carson whose book Multiple Processing Systems for Real-Time Applications was the inspiration for this work, and Jim Gray whose many writings fueled the fire. They and others who have influenced the book include:

Wendy Bartlett at HP
Richard Buckle at Insession
Victor Berutti at Gravic
Ron Byer, Jr., at NetWeave
Gary Chatterton at Sombers
Robert Cline at JPMI
Dick Davis at Gravic
John Dennis at HP
Jim Gray at Microsoft
John Hoffmann at Sombers
Bill Holenstein at Gravic
Denise Holenstein at Gravic
Dan Hoppmann at A. G. Edwards
Ron Hopson at NetWeave

ITUG Connection staff
Clark Jablon at Akin Gump
Bill Knapp at Gravic
Ron LaPedis at HP
Burt Liebowitz, Consultant
Malcolm Mosher at HP
Mike Nee at NTI
Carl Niehaus at HP
Janice Reeder at Sombers
Julie Scherer at Gravic
Paul Siegel at Sombers
Gary Strickler at Gravic
Mark Waterstraat at Insession

Alan Wood at HP

About the Authors

Dr. Wilbur H. (Bill) Highleyman brings more than forty years experience in the design and implementation of computer systems to his position as Chairman of The Sombers Group, Inc. SGI is a turnkey custom software house specializing in the development of real-time, on-line data processing systems, with particular emphasis on fault-tolerant systems and large communications-oriented systems. He is also Chairman of NetWeave Corporation, which developed the middleware product NetWeave. NetWeave is used to integrate heterogeneous computing systems at both the messaging and the database levels. Dr. Highleyman, a graduate of Rensselaer Polytechnic Institute and MIT, earned his doctorate in electrical engineering from Polytechnic Institute of Brooklyn. He has published extensively on availability, performance, middleware, and testing and is the author of "Performance Analysis of Transaction Processing Systems," published by Prentice-Hall. He holds four patents and can be reached at billh@sombers.com.

Paul J. Holenstein is Executive Vice President of Gravic, Inc., the makers of the ITI Shadowbase line of data replication products. Shadowbase is a low latency, high performance real-time data replication engine that provides disaster recovery as well as heterogeneous data transfer. Mr. Holenstein has more than twenty-two years of experience providing architectural designs, implementations, and turnkey application development solutions on a variety of UNIX, Windows, and VMS platforms, with his HP NSK experience dating back to the NonStop I days. He was previously President of Compucon Services Corporation, a turnkey software consultancy. Mr. Holenstein's areas of expertise include high-availability designs, data replication technologies, disaster recovery planning, heterogeneous application and data integration, communications, and performance analysis. Mr. Holenstein, an HP-certified Accredited Systems Engineer (ASE), earned his undergraduate degree in computer engineering from Bucknell University and a master's degree in computer science from Villanova University. He has co-founded two successful companies and holds patents in the field of data replication. He can be reached at shadowbase@iticsc.com.

Dr. Bruce D. Holenstein is President and CEO of Gravic, Inc. Gravic's ITI Shadowbase software supports many of the architectures described in this book and operates on systems such as UNIX, Windows, NonStop and other platforms running databases including

Oracle, Sybase, DB/2, and SQL/MP. Dr. Holenstein began his career in software development in 1980 on a Tandem NonStop I. His fields of expertise include algorithms, mathematical modeling, availability architectures, data replication, pattern recognition systems, process control, and turnkey software. Dr. Holenstein earned his undergraduate degree in Electrical Engineering from Bucknell University and his doctorate from the University of Pennsylvania. Dr. Holenstein has co-founded and run three successful companies and holds patents in the field of data replication. He can be reached at shadowbase@iticsc.com.

Dr. Bill Highleyman
Paul J. Holenstein
Dr. Bruce Holenstein
Paoli, Pennsylvania
July, 2003

Part 1 - Breaking the Four 9s Barrier

Chapter 1 - The 9s Game

True or False:

- *Adding processors to a multi-processor redundant system increases its reliability.*

- *A 16-processor redundant system has the reliability of a UNIX box.*

- *A fault-tolerant processor board is less reliable than a UNIX processor board.*

The above are indeed provocative questions, and their answers are not straightforward. This chapter develops some simple, though not necessarily intuitive, concepts that help answer these and other questions. More importantly, such concepts lead to some straightforward steps that you can take to improve the reliability of your system – steps as simple as being aware of how you allocate processes to processors.

What is Reliability?

The questions above all refer to *reliability*. But before we go much further, we have to agree on how to measure reliability.

There are actually two components that impact the reliability of a system – how long it will work before it fails and then how long it will take to fix. We call the first component the *mean time before failure* (MTBF)[2] and the second *the mean time to repair* (MTR).

To a space satellite designer, reliability is MTBF. Once the satellite fails, it is gone forever. It is not repairable. Clearly, a satellite with a ten-year MTBF is ten times more reliable than a satellite with a one-year MTBF.

However, when it comes to life and property protection, reliability is MTR. In a 911 system, an outage of 30 seconds may be simply an aggravating hiatus; but an outage of one hour can mean death by cardiac arrest or a building burned to the ground.

In large transaction processing systems, reliability is often measured as down time. Down time has a cost associated with it – perhaps $1,000 per hour or $100,000 per hour. (Of course, MTR plays a role here as well; the longer the down time, the higher the cost in many cases – from customer annoyance to lost sales to lost customers.) Down time alternatively can be measured as the proportion of time that a system is up, a measure that we call the *availability* of the system. Since the system is always either up or down, then

$$\text{availability} = A = \frac{\text{MTBF}}{\text{MTBF} + \text{MTR}} \qquad (1\text{-}1a)$$

Note that this also can be written as

$$A = \frac{1}{1 + \frac{\text{MTR}}{\text{MTBF}}} \approx 1 - \frac{\text{MTR}}{\text{MTBF}} \qquad (1\text{-}1b)$$

---

[2] MTBF is often called Mean Time Between Failures. In this sense, it is the average time from one failure to the next failure. However, we use MTBF in the context of the average time from the time that a component is returned to service to the time of its next failure. Therefore, we call MTBF the Mean Time *Before* Failure. Thanks to Dr. Alan Wood of HP for pointing out this subtle but important distinction in semantics.

where "≈" means "approximately equal to," and the approximation is valid so long as MTBF is very much greater than MTR (which certainly is true in the cases that we will be considering).

It is availability that this chapter is all about. When we speak of reliability, we mean availability. A system with .999 availability is more reliable than a system with .99 availability.

More specifically, we will compare the reliability of systems by comparing their probabilities of failure. If a system has an availability of .99, then it has a probability of failure of 1-.99, or .01. That is, it will be down 1% of the time. Using Equation (1-1b),

$$\text{probability of failure} = F = 1 - A \approx \frac{\text{MTR}}{\text{MTBF}} \qquad (1\text{-}2)$$

Jim Gray has characterized availability and reliability in a very folksy way:[3]

"Availability is doing the right thing within the specified response time. Reliability is not doing the wrong thing."

As outlined above, we apply measures to these:

> Availability = Doing the right thing = A
> Failure = Doing the wrong thing = $F = (1-A)$
> Reliability = Not doing the wrong thing = $1/(1-A)$

Consider System A, which is up 99% of the time, and System B, which is up 99.9% of the time. System A has an availability of .99, a failure probability of .01, and a reliability measure of 100. System B has an availability of .999, a failure probability of .001, and a reliability measure of 1000. Thus, we say that System B is ten times more reliable than System A.

---

[3] Gray, J.; "*Why Do Computers Stop and What Can We Do About It?*" 5<u>th</u> <u>Symposium on Reliability in Distributed Software and Database Systems</u>; 1986.

This is how we will use the term *reliability* throughout this chapter.

Some Caveats

There is some algebra used in this chapter to develop availability concepts. About the worst relationship looks like

$$A \approx 1 - f(1-a)^{s+1}$$

If you are algebraically challenged, do not despair. Just skip the math and grasp the concept. The concepts are clearly stated and don't depend upon the math for understanding. Besides, charts and tables are provided so that you can use these relationships without ever breaking out a calculator.

Also, we are interested in developing concepts and rules of thumb. This requires that we take a simplistic view of things. You will be tempted to say, "Yes, but my system does this" or "You haven't considered that." True, but we are taking a 50,000 foot view of things in order to develop some general concepts. Moreover, the concepts developed will allow those of you who are motivated to drop down to a 5,000 foot view. A 500 foot view, however, is probably obscured by a lack of good data and too many trees in your way.

The simplistic view presented in this chapter is most applicable to repairable systems. Software failures generally are recovered rather than repaired. They are more complex and are considered in Chapter 5.

9s – The Measure of Availability

When we calculate availability for today's systems, we will get numbers like .99999. Saying this gets cumbersome and can lose the meaning. So we talk about availability in terms of the number of 9s.

".99999" is "five 9s." ".998" is "a little less than three 9s." ".99992" is "a little more than four 9s."

Though we will speak of 9s, this measure can be converted to average down time over any given period, as shown in Table 1-1.

Nines	% Available	Hours/Year Down Time	Minutes/Month Down Time
2	99%	87.60	438.
3	99.9%	8.76	43.8
4	99.99%	.88	4.38
5	99.999%	.09	.44
6	99.9999%	.01	.04

Average 24x7 Down Time
Table 1-1

Of course, an availability of three 9s does not mean that the system will be down 8.76 hours each year. It means that over a sufficiently long period of time, one can expect that the system will be down an average of eight or nine hours per year. This could occur as short 1 minute failures every 17 hours or as a one-day failure every three years.

Specifically, knowing the availability tells us nothing about the MTBF or MTR. But knowing the availability and either the MTBF or the MTR tells us the other. More to the point, from Equation (1-1b) we can deduce that

$$MTBF \approx MTR/(1-A) \qquad (1\text{-}3)$$
$$MTR \approx MTBF(1-A) \qquad (1\text{-}4)$$

Thus, if we know that our availability is three 9s, and if we have an MTR of 4 hours, then we have an MTBF of 4,000 hours.

Dr. Bill Highleyman, Paul J. Holenstein, and Dr. Bruce Holenstein

Today's Systems

How do today's systems rate so far as availability is concerned? Results compiled by the Gartner Group[4] indicate the following availabilities:

HP NonStop™	.9999
Mainframe	.999
Open VMS	.998
AS400	.998
HPUX	.996
Tru64	.996
Solaris	.995
NT Cluster	.992 - .995

Thus, mainframes are four to five times more reliable than UNIX systems, and HP NonStop systems are ten times more reliable than mainframes.

Simple Systems

Let us start our conceptual journey by looking at the two simplest systems and by reviewing a little probability theory along the way.

We consider a system made up of subsystems, each subsystem with an availability of a. The availability of the entire system is A.

Non-Redundant System

Figure 1-1 shows a system comprising two non-redundant subsystems. Both must work in order for the system to work.

---

[4] Gartner Group; 2002.

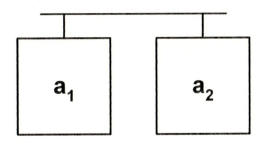

**Non-Redundant System
Figure 1-1**

Let the availability of subsystem 1 be a_1 and of subsystem 2 be a_2. Remember that each of these availabilities is the probability that the subsystem will be operational. In order that the entire system be operational, both subsystem 1 *and* subsystem 2 must be operational. The probability of this is the product of the component probabilities:

$$A = a_1 a_2 \qquad (1\text{-}5)$$

Rule 1: *If all subsystems must be operational, then the availability of the system is the product of the availabilities of the subsystems.*

Redundant System

Figure 1-2 shows a redundant system comprising two identical subsystems, but in this case the system is operational if either subsystem 1 is operational *or* if subsystem 2 is operational. In order for the system to be down, both subsystem 1 *and* subsystem 2 must be down. Since the probability that either subsystem will be down is $(1-a)$, then the probability that both will be down is $(1-a)^2$. The system availability is therefore

$$A = 1 - (1-a)^2 \qquad (1\text{-}6)$$

Dr. Bill Highleyman, Paul J. Holenstein, and Dr. Bruce Holenstein

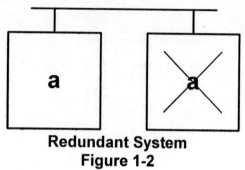

**Redundant System
Figure 1-2**

Double Your 9s

Let us explore Equation (1-6) a little further. If the subsystem availability a is .99, and if we add an additional subsystem so that we have a spare subsystem, then the system availability A of the resulting redundant subsystem is

$$A = 1 - .01 \times .01 = .9999$$

Note that we have doubled the 9s from a subsystem availability of two 9s to a system availability of four 9s. This is the beauty of redundant systems.

Rule 2: *Providing a backup doubles the 9s.*

This is the basis for the high reliability of fault-tolerant systems and for the even higher reliability that can be achieved by replicating a system using data replication techniques (more about that in Chapter 2).

The Real Fault-Tolerant World

Redundant systems are the basis for the high availability (and high scalability as well) of fault-tolerant systems. But these systems are a bit more complex than the simple systems that we have just considered:

- They comprise multiple redundant subsystems – processors, processes, disks, communications.

- Processes critical to system operation are replicated as process pairs.

- Processes are distributed randomly across processors (usually to satisfy load balancing considerations).

Consistent with our 50,000 foot view, we will consider a fault-tolerant system as a single group of like subsystems. Therefore, a subsystem is a processor and its collection of disks and other peripherals. Certainly in a real system, each subsystem will be somewhat different since different processors have associated with them different numbers of devices; but our assumption that all subsystems are similar is warranted by the simplifications that allow us to develop some general concepts.

Furthermore, we will assume a subsystem availability of .995. This is close to the K-series subsystem availability of .996 reported to one of the authors by Tandem in the mid-1990s. Today's systems undoubtedly comprise more reliable components and are manufactured using higher quality techniques, but they are also more complex. It is therefore assumed that a subsystem availability of .995 is still in the ballpark.

Note that this value for availability includes all sources of failure: hardware, software, maintenance, and operations. More about that in Chapter 5.

Randomly Distributed Process Pairs

So far as availability is concerned, the heart of a fault-tolerant system is its critical processes. The loss of any one of these processes will cause a system failure, either immediately or after a short period of time due to system degradation. For instance, in HP NonStop servers, critical processes include the disk processes (DP2), PATHMON, and a slew of monitors for communication and other subsystems.

Therefore, these processes are provided as process pairs so that they will survive any single processor failure. Coupled with transaction protection that guarantees that no data will be corrupted as a result of a fault, these features provide the high availability for which fault-tolerant systems are known.

But a dual processor failure may take down a critical process pair and result in a system outage. Let us take a look at a four processor system in which critical processes are randomly distributed across all processors so that any dual processor failure will take down the system (Figure 1-3).

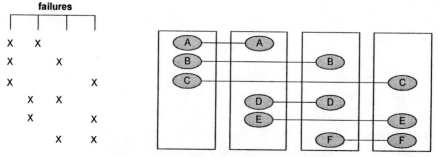

Any pair of processor failures causes a system failure

Randomly Distributed Processes
Figure 1-3

Note that there are six possible ways that two out of four processors can fail. We call these *failure modes*. Since any given two processors will fail with a probability of $(1-a)^2$, and since there are six ways that this can happen, then the system will fail with a probability of $6(1-a)^2$. Thus, its availability is[5]

---

[5] This relation is an approximation since it does not account for failure modes involving more than two subsystems. Since the probability of three or more failures is extremely small, the relation is quite accurate. Besides, it's a lot simpler than the fully accurate relation.

Breaking the Availability Barrier

$$A \approx 1 - 6(1-a)^2$$

You can probably figure out that for *n* processors, the number of failure modes is $n(n-1)/2$. For the above example, $n=4$. The number of failure modes is 4x3/2=6.

Process/Processor Pairing

Figure 1-4 shows an alternate strategy for distributing process pairs. Processors are organized into pairs, and process pairs are constrained to run only in processor pairs. For a four-processor system, there are only two failure modes. Either the first pair of processors must fail or the second pair must fail in order to cause a system failure. Thus, the availability of this configuration is

$$A \approx 1 - 2(1-a)^2$$

This configuration has three times the reliability of the randomly distributed configuration. In general, for *n* subsystems, the number of failure modes is *n/2* for this strategy.

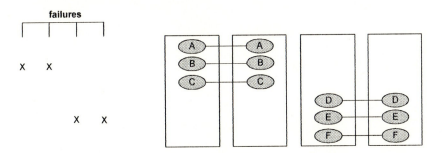

Only certain pairs of processor failures cause a system failure

**Process Pairing
Figure 1-4**

Dr. Bill Highleyman, Paul J. Holenstein, and Dr. Bruce Holenstein

In fact, the advantage of process/processor pairing gets better as the system gets larger. Consider an eight-processor system (where f is the number of failure modes):

	n	**a**	**f**	**A**
random	8	.995	28	.9993
paired	8	.995	4	.9999

To review the terminology in the above table,

> n is the number of subsystems in the system,
> a is the subsystem availability,
> f is the number of failure modes, and
> A is the system availability.

For an 8-processor system, a paired configuration is seven times as reliable as a random configuration. For a 16-processor system, reliability is improved by a factor of fifteen!

Rule 3: *System reliability is inversely proportional to the number of failure modes.*

Rule 4: *Organize processors into pairs, and allocate each process pair only to a processor pair.*

Availability in General

We have seen above that failure probability is proportional to the number of failure modes. Thus, letting f be the number of failure modes, we may write

$$A \approx 1 - f(1-a)^2$$

This relationship assumes that each process is backed up by only one other process (we can consider the backup process as a spare which is put into service if the primary process fails).

But what if we have two spares? Then any given failure occurs only with a probability of $(1-a)^3$. In general, if we have s spares, then the system will fail only if we have $s+1$ subsystem failures. Any particular failure of $s+1$ subsystems will occur with a probability of $(1-a)^{s+1}$, and the system availability becomes

$$A = 1-F \approx 1-f(1-a)^{s+1} \qquad (1\text{-}7)$$

This is our general availability equation.[6] Note that it reduces to our simple example represented by Equation (1-6) for $f=1$ and $s=1$.[7] This result can be expressed as

> **Rule 5:** *If a system can withstand the failure of s subsystems, then the probability of failure of the system is the product of the probability of failures of (s+1) systems.*

There is one assumption that is inherent in our discussion so far, and that is that the system is returned to service as soon as a failed subsystem has been repaired. It needs no further recovery. This assumption is explored further in Chapter 5, *"The Facts of Life."*

More Sparing

Note that from Equation (1-7), reliability increases exponentially with the number of spares. If a is .99, each additional level of sparing adds another two 9s to the system availability. Single sparing gives a system availability of four 9s, double sparing gives an availability of six 9s, and so on.

---

[6] As we said earlier, this is an approximation. However, not only is it conservative in that it gives a lower value for A than the actual value, but it is within 5% for the range of values in which we are interested. See Appendix 2 for more on this.

[7] This result is an extension of an excellent summary of availability found in Chapter 8, *"Reliability Calculations,"* Multiple Processing Systems for Real Time Applications by Burt H. Liebowitz and John H. Carson, Prentice-Hall;1985.

Rule 6: *System availability increases dramatically with increased sparing. Each additional level of sparing adds a subsystem's worth of 9s to the overall system availability.*

How do we increase process sparing in fault-tolerant systems? There are several methods:

- For process pairs (i.e., a primary process with a backup process waiting to take over in the event of a primary process failure), allow a process to start a new backup in a surviving processor if the process loses its backup due to a processor failure.

- For persistent processes that are restarted in another processor by a monitor, give the monitor the choice of more than two processors in which to restart the process (of course, the monitor must be redundant as well).

- Design the application to allow more than one active process or more than one backup process.

How Many Failure Modes?

The worst case for availability is a random distribution of processes, since in this case any pair of processor failures will take the system down. For n subsystems and s spares, the number of failure modes for this case is the number of ways that $s+1$ subsystems can fail out of n systems.[8] These maximum values for f are shown in Table 1-2.

We can see from this table that the maximum number of failure modes for a single-spared 16-processor system is 120. However, we know that if we pair processors and processes, we can reduce the

---

[8] For Math Nuts: This is $n!/(n-s-1)!(s+1)!$

number of failure modes to eight, a 15:1 reduction as we have earlier noted.

Since reliability is inversely proportional to the number of failure modes (Rule 3), we can lose more than a nine from our achievable availability for a 16-processor system if we are not careful with process allocation. More about this later.

Spares (s)	Processors (n)							
	2	4	6	8	10	12	14	16
0	2	4	6	8	10	12	14	16
1	1	6	15	28	45	66	91	120
2		4	20	56	120	220	364	560
3		1	15	70	210	495	1001	1820
4			6	56	252	792	2002	4368
5			1	28	210	924	3003	8008
6				8	120	792	3432	11440
7				1	45	495	3003	12870
8					10	220	2002	11440
9					1	66	1001	8008
10						12	364	4368
11						1	91	1820
12							14	560
13							1	120
14								16
15								1

Maximum Failure Modes (f)
Table 1-2

The Impact of Repair Time

So far, we have talked about availability as the predominant measure of reliability. But as we indicated in the opening to this chapter, the system mean time to repair, MTR, is often an equally important parameter. Let us look at system MTR and its relation to subsystem mean time to repair, mtr.

From Equation (1-1a), we can express subsystem availability a in terms of its *mtbf* and *mtr*:

$$a = \frac{mtbf}{mtbf + mtr}$$

where
- a is the subsystem availability.
- *mtbf* is the subsystem mean time before failure.
- *mtr* is the subsystem mean time to repair.

Note that we are using upper case MTBF and MTR to represent the system and lower case mtbf and mtr to represent the subsystem.

A little algebraic manipulation results in

$$1 - a = 1 - \frac{1}{1 + \frac{mtr}{mtbf}} \approx \frac{mtr}{mtbf} \qquad (1\text{-}8)$$

The approximation depends upon mtbf being much greater than mtr. This is certainly true by orders of magnitude in the systems that we are considering.

Substituting Equation (1-8) into Equation (1-7), we have

$$A = 1 - F \approx 1 - f\left(\frac{mtr}{mtbf}\right)^{s+1} \qquad (1\text{-}9)$$

We see that reliability is exponentially affected by subsystem mtr. For one spare ($s = 1$), the system failure probability will be cut by a factor of four if we can cut subsystem mtr in half.

But how does subsystem mtr affect the overall system MTR and its MTBF? It can be shown[9] that

---

[9] If $(s+1)$ subsystems are being repaired independently, and if each requires an average time of *mtr* to repair, then they are being repaired at a rate of $(s+1)/mtr$.

$$\text{MTR} = \frac{mtr}{(s+1)} \qquad (1\text{-}10)$$

For one spare, system MTR is half the subsystem mtr. Thus, for one spare, if our subsystem mtr is four hours, then our system MTR is two hours. (This assumes that the repairs of multiple subsystems are independent of each other.)

> **Rule 7:** *For a single spare system, the system MTR is one-half the subsystem mtr.*

Furthermore, if we reduce mtr by a factor of k, we will reduce MTR by a factor of k. Since we have seen that the failure probability will be reduced by k^2 (Equation (1-9)), then from Equation (1-2) we can conclude that we will increase our system MTBF by a factor of k.

> **Rule 8:** *For the case of a single spare, cutting subsystem mtr by a factor of k will reduce system MTR by a factor of k and increase the system MTBF by a factor of k, thus increasing system reliability by a factor of k^2.*

For instance, let us say that our system has an MTBF of five years and an mtr of four hours, leading to an MTR of two hours. If we can cut mtr in half to two hours, our system MTR will be reduced to one hour; and our system MTBF will be increased to ten years. Our reliability has increased by a factor of four, as indicated above.

Some Helpful Charts

Figure 1-5 shows availability as a function of the number of processors and the number of spares for random distribution of processes. Note that no matter the number of processors, each additional spare adds about two 9s to the system availability.

Thus, the average time to the next repair, which will return the system to service, is $mtr/(s+1)$.

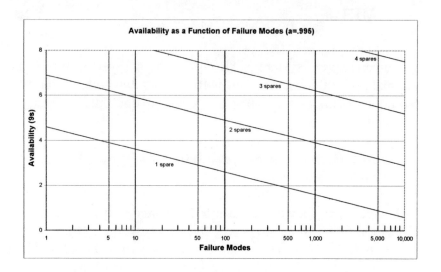

**Availability as a Function of Processors and Spares
Figure 1-5**

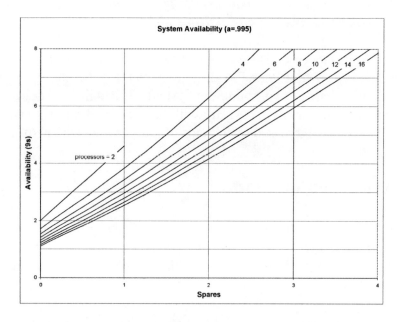

**Availability as a Function of Spares, Failure Modes
Figure 1-6**

Breaking the Availability Barrier

Figure 1-6 shows availability as a function of spares and failure modes. Let's look at process pairing and random distribution for 16 processors. For process pairing (f=8), system availability is almost four 9s. For random distribution (f=120), system availability is about two and a half nines. My! That's about the availability of a UNIX box.

Answers

Let's return to our initial provocative statements.

- *Adding processors to a multi-processor redundant system increases reliability.*

Well, it all depends. If sparing remains the same, then the number of failure modes increases; and reliability decreases. However, if the extra processors are used to increase sparing, then reliability can dramatically increase. However, since we don't generally change the design of the system when we add processors, this statement is typically false. Adding processors reduces reliability.

- *A 16-processor redundant system has the reliability of a UNIX box.*

Again, it all depends. If you are distributing processes randomly, then this is true. However, if you are intelligent in the way you distribute processes, your system reliability will beat that of a UNIX system by an order of magnitude or more. So hopefully, your answer to this is false.

- *A fault-tolerant processor board is less reliable than a UNIX processor board.*

The answer to this one is probably true, but so what? In order to build a truly fault-tolerant system, all components, both hardware and software, must be fail-fast so that corrupted data does not get propagated. To achieve this, fault-tolerant processor boards often use a pair of on-board, lock-step processors which continually compare results. If there is a mismatch, the processor board shuts down immediately. Thus, the processor board can fail if either on-board processor fails, giving two failure modes instead of one for the UNIX processor. Given comparable component and manufacturing quality and comparable component count per processor, we expect the fault-tolerant processor board to fail twice as often as a UNIX processor. But this is a trivial price to pay for the ultimate fault tolerance provided. A fault-tolerant architecture beats non-fault-tolerant architecture by two 9s or so (100 times more reliable).

Summary

We summarize the concepts presented above by considering how we might improve reliability. Our general availability relationships of Equations (1-1), (1-7), (1-9), and (1-10) state that[10]

$$A = \frac{MTBF}{MTBF+MTR} \approx 1 - \frac{MTR}{MTBF} \qquad (1\text{-}1a,b)$$

$$A \approx 1 - f(1-a)^{s+1} \approx 1 - f\left(\frac{mtr}{mtbf}\right)^{s+1} \qquad (1\text{-}7), (1\text{-}9)$$

$$MTR = \frac{mtr}{s+1} \qquad (1\text{-}10)$$

From the above equations, we can also determine that

$$MTBF \approx \frac{mtbf}{f(s+1)}\left(\frac{mtbf}{mtr}\right)^{s} \qquad (1\text{-}11)$$

---

[10] These equations are restatements of the Einhorn relationships. See Einhorn, S. J.; "*Reliability Prediction for Repairable Redundant Systems*," Proceedings of the IEEE; February, 1963.

Breaking the Availability Barrier

These expressions relate system availability, A, system mean time to repair, MTR, and system mean time before failure, MTBF, to four parameters around which we can get our hands – f, s, $mtbf$, and mtr:

    mtbf   Subsystem mean time before failure is out of our control. Not much we can do about that.

    mtr   Reductions in subsystem repair time have an exponential impact on availability. In a one-spare system, cutting subsystem repair time in half provides a four-fold improvement in reliability. Cutting it by a factor of ten provides 100 times more reliability – two 9s on the availability scale. Consider a tighter service level agreement (SLA) or, for larger users, on-site spares and on-site maintenance.

    s   There's not much that we as users can do to increase sparing of manufacturer-supplied critical processes. The fault-tolerant system vendor must decide that the significant software development effort to do this is worthwhile. Until this step is taken, there is not much sense in critical application processes being written with sparing in excess of one.

    f   Ah! We can control the number of failure modes. As we have shown, the intelligent distribution of critical processes can reduce failure rates significantly. If we work at it, we actually can pick up one or two 9s.

How Far Should We Go?

From Figure 1-6, we see that we can achieve an availability of four 9s with our fault-tolerant systems if we can keep the failure modes to five or less. Isn't this enough?

Dr. Bill Highleyman, Paul J. Holenstein, and Dr. Bruce Holenstein

The Standish Group[11] has defined the following application categories and their required availability:

Class	9s
Non-critical	2
Task critical	3
Business critical	4
Mission critical	5
Safety critical	6

**Availability Requirements
(The Standish Group)
Table 1-3**

So if you have a need that you might characterize as business critical, mission-critical, or safety-critical, you had better mind your 9s.

A Case Study

Amtrak provided a real-life case study of the above concepts with their real-time train control system for the busy Northeast Corridor. Shown at a very high level in Figure 1-7, Amtrak's system uses a NonStop system to monitor and control trains via a duplexed communication link to track-side devices (signals, switches, occupancy detectors). A redundant console system is used by the train dispatchers to monitor and direct train traffic.

A detailed analysis of the system availability (whose results are shown in Figure 1-7) predicted a system availability of .9995. Not bad! Unfortunately, the specifications for the system required an availability of .9998. The design had missed the reliability mark by a factor of 2.5. To make matters worse, this was a fixed price contract; and the system was not going to be accepted unless the availability requirement was met.

---

[11] Standish Group; 2002.

Breaking the Availability Barrier

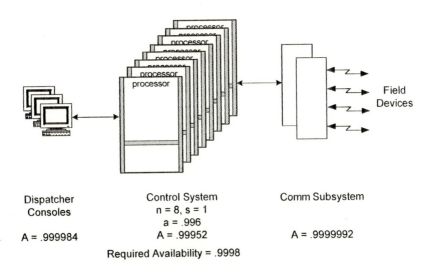

**Amtrak Train Control System
Figure 1-7**

It was clear that the culprit was the eight-processor K-series system that had an availability of .99952 based on the worst case of 28 failure modes. However, a little algebra showed that if the number of failure modes were reduced to 12, the specification would be met. Process allocation guidelines were put in place, and the system was accepted.

This was an inexpensive fix to a potentially very expensive problem and used the concepts discussed in this chapter.

What's Next?

Bear in mind that all we have really talked about so far are general availability concepts as applied to redundant hardware systems. The consideration of faults caused by software and human errors is a much more complex subject which is considered in Chapter 5, *"The Facts of Life."*

In Chapter 2, *"Splitting Systems,"* we explore in more depth the availability considerations and advantages of replicating and splitting systems.

Chapter 2 - System Splitting

In Chapter 1, we talked about the dramatic increase that can be achieved in system availability by making a system redundant so that it is capable of tolerating one or more failures. This, of course, is the basis for fault-tolerant systems, in which availabilities of four 9s are achievable (that is, the system will be up 99.99% of the time).

However, some applications require much more reliability than this. To achieve such high availability, a common technique is to provide a secondary system to back up the primary system. As we showed in Chapter 1, providing a single level of redundancy doubles the nines. Thus, redundant systems can potentially achieve availabilities in the order of eight 9s.

In some applications, it may be unnecessary to replicate the entire system. For instance, the system can be split into two independent half systems and provide the same enhanced availability at little additional cost.

This chapter considers some of the availability advantages associated with replicating and splitting systems.

The Availability Relation

As a quick review, the availability of a redundant system is given by

$$A = 1 - F \approx 1 - f(1-a)^{s+1} \tag{2-1}$$

where it is assumed that the system can be restored to service as soon as a failed subsystem is repaired and where

- A is the system availability (the portion of time that it is operational).
- F is the probability that the system will be down.
- a is the availability of a subsystem.
- s is the sparing level (i.e., $s+1$ subsystems must fail in order for the system to fail).
- f is the number of failure modes or the number of ways that $s+1$ subsystem failures will cause a system outage.

The approximation in Equation (2-1) assumes that the subsystem failure probability $(1-a)$ is very much less than one.

The rationale for this relationship is simple. $(1-a)$ is the probability that a subsystem will fail. $(1-a)^{s+1}$ is the probability that $(s+1)$ subsystems will fail. Since this can happen in f ways that will take the system down, then the probability that the system will be down is $f(1-a)^{s+1}$. System availability is one minus the probability that the system will be down.

The number of failure modes is an aspect that can be controlled by proper system design and configuration, as discussed in Chapter 1. The worst case is that in which any combination of $s+1$ subsystems will cause a system outage. In this case, f is the number of ways one can select $(s+1)$ subsystems from the total of n subsystems making up the system:

$$f = \binom{n}{s+1} = \frac{n!}{(s+1)!(n-s-1)!} \tag{2-2a}$$

For the singly redundant case, $s=1$; and the number of failure modes f is

$$f = \frac{n(n-1)}{2} \tag{2-2b}$$

where *n* is the number of processors in the system.

Full Replication

A very common way to protect the availability of a system is to fully replicate it, often in a different geographical region. The systems may be configured as a primary system and as a backup system, in which case the primary system handles the full load and maintains a copy of its database at the backup site via data replication (Figure 2-1a). This often is called an active/passive architecture. Alternatively, both systems may share the load using a variety of strategies described later. In the case of load sharing, each system will keep its companion's database synchronized via data replication (Figure 2-1b). In either case, provision is made for the surviving system to assume the entire load in the event of a system failure. This is called an active/active architecture.

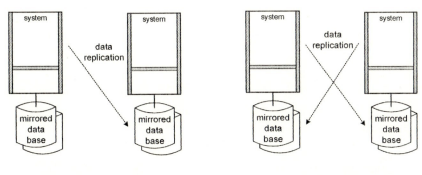

a) Primary/Backup
(active/passive)

b) Load Sharing
(active/active)

Full Replication
Figure 2-1

We know intuitively that system replication can dramatically improve reliability. But let us look at this quantitatively.

In the case of full replication, there is an overall system failure only if both systems fail. If the probability of failure of a single system is F_1, then the probability that both systems will be down, F, is

Dr. Bill Highleyman, Paul J. Holenstein, and Dr. Bruce Holenstein

$$F = F_1^2 \qquad (2\text{-}3)$$

The overall system availability A is the probability that both systems will not be down:

$$A = 1 - F_1^2 \qquad (2\text{-}4)$$

An important implication of Equations (2-3) and (2-4) is shown in the following table, which calculates overall availability A as a function of single system availability A_1:

Single System Availability A_1	Single System Failure Probability $F_1 = 1 - A_1$	Replicated System Availability $A = 1 - F_1^2$
.9	.1	.99
.99	.01	.9999
.999	.001	.999999

We see here that *full replication doubles the 9s* (see Rule 2 in Chapter 1).

An insight into this power of replication can be gained by looking at our approximation to availability given by Equation (2-1). According to this relationship, the probability of failure for a single system, F_1, is approximately

$$F_1 \approx f(1-a)^{s+1}$$

According to Equation (2-3), the probability of failure for the replicated system, F, is then

$$F = F_1^2 \approx f^2(1-a)^{2(s+1)}$$

In effect, the impact of sparing has been doubled due to the doubling of the exponent for the (1-a) term. If it takes two failures to cause a system failure in one of the systems, it will take *four* failures to cause a replicated system failure. True, the number of failure modes has increased (from f to $f^{\,2}$); but this is generally small compared to the effect of the extra sparing.

System Splitting

Simple Splitting

Rather than fully replicating a system at twice the cost of a single system, let us consider splitting a system into two equal half pieces, or nodes, perhaps geographically separated, as shown in Figure 2-2. A fully replicated system can either be an active/active or an active/passive system, as described in the previous section. However, the split system is necessarily an active/active system. Each node carries its share of the load during normal operation and keeps its companion's database in synchronism via data replication. However, should one node fail, the other will assume the full load. There are, of course, many considerations involved in implementing this sort of solution, such as load shedding in the event of reduced capacity and the switching of users to the surviving system. Assuming that these problems are solvable, let us look at the availability implications of this solution.

For purposes of illustration, let us assume that a system or a node is configured with one spare element (s=1) and that any dual failure will bring down that system or node. Then for a system with n elements, the number of failure modes f is, from Equation (2-2b),

$$f = \frac{n(n-1)}{2}$$

and the failure probability for that system is

$$F \approx \frac{n(n-1)}{2}(1-a)^2 \qquad (2\text{-}5)$$

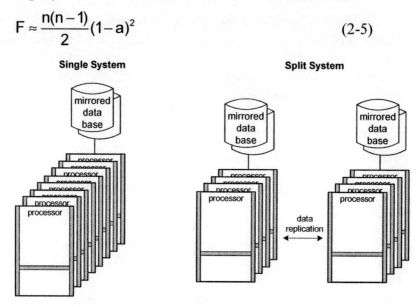

Splitting a System
Figure 2-2

Referring to Figure 2-2, if we split this system into two equal nodes of $n/2$ processors each, then each node will have a failure rate F_n of

$$F_n \approx \frac{\frac{n}{2}\left(\frac{n}{2}-1\right)}{2}(1-a)^2 \qquad (2\text{-}6)$$

where F_n is the failure probability of a single node in the split system. The split system will maintain full capacity so long as both of its nodes are operating. The probability that either node is not operating, assuming that this probability is small, is

$$F_{100} \approx 2F_n \approx \frac{\frac{n}{2}(n-2)}{2}(1-a)^2 \qquad (2\text{-}7)$$

where F_{100} denotes the probability of the split system failing to provide 100% capacity.

Let us define a reliability ratio that is the ratio of the failure rate, F, of the full system (Equation (2-5)) to that of the split system, F_{100} (Equation (2-7)), where failure means the failure to provide 100% capacity. Then

$$\frac{F}{F_{100}} \approx \frac{\frac{n(n-1)}{2}}{\frac{n}{2}(n-2)} = 2\frac{n-1}{n-2} > 2 \qquad (2\text{-}8)$$

Since this reliability ratio always is greater than 2, then we can conclude that the split system will provide full 100% capacity with a reliability at least twice as great as a single system providing the same capacity.

This is a direct result of the reduction in failure modes. For instance, an eight- processor system has 28 failure modes for a single spare configuration (8x7/2 from Equation (2-2b)). If we split this 8-processor system into two 4-processor nodes, each of these nodes will have six failure modes (4x3/2). Since the failure of either node constitutes a failure to provide 100% capacity, then the total failure modes for the split system is 2x6, or 12. This is less than half of the 28 failure modes for the full system.

In addition, the split system will provide dramatically better availability if 50% capacity is acceptable. In this case, the availability of the system will be double the 9s of one of the nodes (which is already significantly better than the original un-replicated system).[12]

---

[12] Note: There are patents pending on the various methods for achieving the advantages of system splitting.

By way of example, consider an 8-processor system that is split into two 4-processor nodes.[13] All system elements have an availability of .995, and each node is configured to have one spare. The relative system availabilities are as follows:

Configuration	Availability of 100% Capacity	Availability of 50% Capacity
Single System	.9993	---
Split System	.9997	.99999991

As we can see,

- the split system provides 100% capacity more reliably than the single system (by a failure improvement of more than a factor of 2).

- the split system provides some capacity with an availability dramatically better than that of the single system.

Multiple Splitting

The concept of splitting a system into two nodes can be extended to splitting it into k nodes. For instance, one can consider splitting a 16-processor system into four geographically dispersed 4-processor nodes.

---

[13] An interesting variant of system splitting is found in the Itanium-based HP NonStop server. To provide significantly improved processor reliability, HP split its tightly-coupled RISC microprocessors used to check each other into two independent microprocessors. By doing so, they reduced the number of failure modes by a factor of 4 and increased the sparing level from one to three. See R. Buckle, W. Highleyman, "*The New NonStop Advanced Architecture: A Massive Jump in Processor Reliability,*" The Connection, Issue 24, No. 5; September/October, 2003.

Breaking the Availability Barrier

Assuming that any dual processor failure will take down a node, then node failure rate F_n (given for $k=2$ in Equation (2-6)) is

$$F_n \approx \frac{\frac{n}{k}\left(\frac{n}{k}-1\right)}{2}(1-a)^2 \qquad (2-9)$$

Since partial capacity can be lost in k ways (any one of the k nodes failing), then the probability that the system will not provide 100% capacity, F_{100}, is

$$F_{100} \approx kF_n = \frac{\frac{n}{k}(n-k)}{2}(1-a)^2 \qquad (2-10)$$

The reliability ratio is then

$$\frac{F}{F_{100}} \approx \frac{\frac{n(n-1)}{2}}{\frac{n}{k}(n-k)} = k\frac{n-1}{n-k} > k \qquad (2-11)$$

Thus,

- The split system will provide 100% capacity at least k times more reliably than the single system (in the sense that its failure probability is reduced by a factor of more than k compared to the normal system.

- If the split system does suffer an outage, it loses only 1/k of its capacity (for instance, 25% for a four-way split).

Rules 1 through 8 were given in Chapter 1. Here are some additional ones:

Rule 9: *If a system is split into k parts, the resulting system network will be more than k times as reliable as the original* system and still will deliver (k-1)/k of the system capacity in the event of an outage.

Rule 10: *If a system is split into k parts, the chance of losing more than 1/k of its capacity is many, many times less than the chance that the single system will lose all of its capacity.*

Impact on Mean Time Before Failure

In Chapter 1, we showed that mean time before failure, MTBF, mean time to repair, MTR, and the system failure probability, F, were approximately related as follows:

$$MTBF \approx \frac{MTR}{F}$$

That is, MTBF is inversely proportional to the system failure rate. If we cut the failure rate in half, we double the MTBF, all other things remaining equal.

Let us consider separating a 16-processor system into four 4-processor nodes. From Equation (2-11), we see that this increases our reliability (decreases our failure probability) by a factor of 5 (i.e., 4 x 15/12). Thus, if the single system had a mean time before failure (MTBF) of 10 years, we have just extended that to 50 years. Furthermore, in the event of an outage, only 25% of the system capacity is lost. The average time before losing more than 25% capacity is measured in centuries.

In the extreme, splitting a 16-processor system into eight 2-processor nodes will provide a 15-fold availability advantage. Our 10-year system MTBF has now been extended to 150 years, and an outage costs us just 12.5% of capacity.

Of course, splitting a system too finely creates another problem that must be avoided. If a node has been made so small that it cannot handle any meaningful load should it lose one processor, then a single processor failure will, in effect, bring down the node. In this case, the

Breaking the Availability Barrier

node is no longer fault-tolerant. It can be shown that in this case, the benefits of system splitting disappear.

Elimination of Planned Down Time

In addition to the additional fault tolerance achieved by splitting a system into independent nodes, one now has the ability to make hardware and software upgrades one node at a time, thus eliminating planned down time. This is a cornerstone of HP's NonStop Indestructible Scalable Computing initiative[14].

Replication of Data

When we split a system into two or more nodes, we need to replicate the critical components of the database if we are to achieve the benefits of system splitting. Should we split a system by a factor greater than two – say by a factor of k – we do not have to replicate the database k times. We do have to make sure, however, that it is replicated at least once so that the entire database is available even in the presence of a node failure. For instance, the database can be partitioned so that each node in the network has $2/k$ of the database connected to it, as shown in Figure 2-3. Each of these replicated database partitions must be kept in synchronization via some sort of data replication.

There are several options as well as some serious issues involved with the replication of data across systems.

- The simplest architecture is the fully replicated primary/backup (or active/passive) system described earlier (see Figure 2-1a). In this case, the backup database is kept synchronized with that of the primary; but there is no update activity to the database by the backup system (it may, however, be used for read-only operations such as query, reporting, etc.). Here, the only issue is replication latency. Replication latency is the time that it takes to propagate a

---

[14] Bartlett, W.; *"Indestructible Scalable Computing,"* ITUG Summit presentation; September, 2001.

database change from the source database to the target database. Should the primary go down, any updates that are in the replication pipeline would be lost.

- Perhaps the database can be partitioned so that each partition is "owned" by a node, as shown in Figure 2-3. Only that node may make updates to its partition. These updates then are replicated to other nodes for information purposes only. Thus, even in the event of a node failure, all current data is available to all users. If ownership of a partition moves to a surviving node, then full functionality is maintained even in the event of a node failure. Again, the primary data replication concern is lost updates due to replication latency.

- In the general case, any user at any node can update any data item. This is the active/active configuration shown in Figure 2-1b. There are several important data replication issues with which to contend when using this approach:

  a) There must be a mechanism to guard against *ping-ponging*, or the replication of an item back to the source of that item.[15]

  b) To the extent that there is latency in the data replication process, there is a chance that two different updates will be made to the same data item at the same time and that these conflicting updates will be replicated across all systems. These conflicts are called *data collisions*. The database contamination caused by data collisions must first be detected and then either be corrected manually or by automatic conflict resolution via business rules. Data collision detection and resolution are discussed in some detail in Chapter 3, "*Asynchronous Replication.*"

  c) If the distributed databases are tightly synchronized to eliminate latency so as to ensure that there will be no data

---

[15] Strickler, G.; et al.; "*Bi-directional Database Replication Scheme for Controlling Ping-Ponging*," United States Patent 6,122,630; Sept. 19, 2000.

conflicts, then system performance may suffer.[16] Synchronous replication performance will be explored further in Chapter 4, "*Synchronous Replication.*"

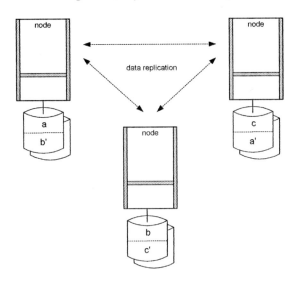

**Database Partitioning Across Nodes
Figure 2-3**

In addition, the data replication facility must be aware of the current network configuration so that it can replicate around node and network faults. Furthermore, provision must be made to switch users from a failed node to an operational node.

Must We Replicate the Database?

We have talked about splitting a system into two or more parts at little additional cost except for a replicate of the database. However, in many large systems, the cost of the disk subsystems can be 70% to 90% of the system cost. It doesn't seem that splitting a system and

---

[16] There are patents pending on various methods for conflict identification, resolution and avoidance. See www.uspto.gov.

replicating the database is the cost effective solution that we suggested.

The solution to this quandary is to note that up until now, we have assumed that the database replicates are mirrored (see Figures 2-1 through 2-3). Is this really necessary? If we are replicating data across the nodes anyway, why does it need to be replicated once again by mirroring the disks at each node?

There is a strong argument that suggests that the database replicates in a split system do not need to be mirrored. It goes as follows:

In 1985, Jim Gray[17] pointed out that the MTBF of disk drives was about 10,000 hours. By 1988, this had grown to 75,000 hours. The year 1992 brought 175,000 hour MTBFs, and by 1996 disk drive MTBFs were at 500,000 hours[18]. Surprisingly, not only disk speed and capacity but also disk availability have been following Moore's Law that performance doubles every eighteen months.

Let us conservatively use a disk MTBF of 100,000 hours. Assuming a leisurely 24-hour repair time, a mirrored disk pair will have an availability of .99999994, or over seven 9s. In fact, since a dual disk failure has a mean time to repair of one-half the disk unit mean time to repair, or 12 hours (see Chapter 1 – *"The 9s Game"*), this means that the MTBF for a mirrored disk pair is almost 240 centuries!

Since the MTBF of today's fault-tolerant systems such as HP NonStop servers is in the order of 10 years, it seems reasonable that database replicates need not be mirrored. It is important only to make sure that there are at least two copies of the data independently accessible somewhere in the network.

---

[17] Gray, J.; *"Why Do Computers Stop and What Can We Do About It?"* 5th Symposium on Reliability in Distributed Software and Database Systems; 1986.

[18] Wong, B. L.; Configuration and Capacity Planning for Solaris Servers, Prentice-Hall.

Breaking the Availability Barrier

Of course, if we have d mirrored disk volumes, then the average time between failures is $240/d$ centuries. However, unless d is very large, the replicated disk farm will still have an MTBF much longer than a processing node.

This leads to two strategies for providing replicated data to all nodes in a split system. One strategy is to split the mirrors (Figure 2-4a). In essence, the database on each node is not mirrored. Mirroring is provided by having a single copy of the database on each of two nodes and by keeping them synchronized via data replication.

Split System Networked Database

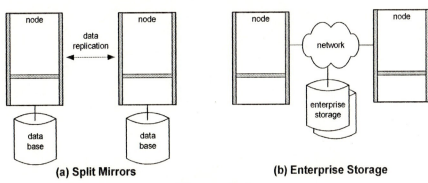

(a) Split Mirrors   (b) Enterprise Storage

Figure 2-4

The other is to provide mirrored enterprise storage accessible to all nodes (Figure 2-4b). Note that split mirrors satisfy the needs for disaster recovery since the database replicates can be geographically separate. Enterprise storage does not satisfy this requirement unless the mirrors can be geographically separate and accessible via different network paths.

Either of these solutions leads us to the desired result – increased availability via system splitting with little if any additional cost.

By the way, you might ask how a mechanical dual disk system can be so much more reliable than dual processors? The answer is simple – no software and no humans. More about this in Chapter 5.

Dr. Bill Highleyman, Paul J. Holenstein, and Dr. Bruce Holenstein

Summary

No advantage comes for free. The significant availability advantages brought about by system splitting have their potential costs. In principle, we can split a system into two or more nodes with little additional hardware cost, we can add data replication to keep them in sync, and we can gain dramatically improved reliability. As we have seen,

- Replicating a system doubles its 9s.

- Splitting a system into k nodes reduces the chance of not providing 100% capacity by at least a factor of k.

- Splitting a system into two or more nodes dramatically increases the availability of at least some processing capacity. In fact, if a system is split into k nodes, the chance of losing more than $1/k$ of its capacity is *many, many* times less than the chance that the single system will lose *all* of its capacity.

- Splitting a system allows us to eliminate planned outages.

However:

- Data replication latency can cause lost updates in the event of a failure.

- Data replication latency also can cause database corruption via conflicting updates if all nodes are allowed to actively process inputs independently without concern for update activity at other nodes.

- Tight database integration to eliminate latency and data collisions may have a negative performance impact.

- Software licensing costs may increase.

- We will see in Chapter 5 – *"The Facts of Life"* that the chance of failover faults has the potential for eroding the availability advantages provided by system splitting.

The good news is that solutions are emerging for these problems, as evidenced by patent activity, published papers, and new product enhancements. Replicating full or partial systems using data replication as the synchronizing mechanism is a powerful method for significantly improving system availability in the presence of system faults, natural disasters, and human errors.

System Splitting and Process Pairing

In Chapter 1, we noted that the availability of a system could be significantly enhanced by the simple technique of carefully allocating process pairs to processors so as to reduce the number of failure modes. In the extreme, for instance, the 120 random failure modes of a 16-processor system could be reduced to only eight failure modes by assigning process pairs to processor pairs. Isn't this the same as splitting a 16-processor system into eight separate two-processor systems, but with no costs of additional data storage and with no problems associated with keeping multiple copies of the database synchronized across the network?

No. Reducing failure modes by allocating process pairs to subsets of processors achieves only part of what is achieved by system splitting. True, the number of failure modes is decreased in similar ways by both techniques. However, the difference is in the sparing. With process allocation, the loss of a critical processor pair means the loss of the entire system. However, if the system instead is split into completely independent nodes, then the loss of a critical processor pair results only in the loss of a node. The other nodes continue to provide full functionality.

For example, if critical process pairs in a 16-processor system were each to be assigned to one of four 4-processor subsets, the number of failure modes would be reduced, but the loss of a critical processor pair would mean the loss of all capacity. If the system instead were to be split into four independent nodes, there would be an equivalent reduction in the number of failure modes. However, the loss of a critical processor pair would result only in the loss of one node, or of 25% of the capacity. The loss of more capacity than this would be almost never.

What's Next?

In this chapter, we have talked about replicating and splitting systems. The replication of an application over several independent and cooperating nodes can dramatically improve the overall availability of the application. However, in order to achieve this availability improvement, it is imperative that, following a nodal failure, the applications on the surviving nodes still have access to the application's database. What is necessarily implied is that the database be itself replicated across two or more nodes.

If there are multiple copies of the database in the application network, then the copies must somehow be kept in synchronization. One way to do this is via asynchronous data replication. Techniques for asynchronous replication and its advantages and disadvantages are discussed in Chapter 3, *"Asynchronous Replication."*

Chapter 3 - Asynchronous Replication

In Chapter 2, "*System Splitting*," we showed that distributing an application over several independent nodes can significantly increase the availability of the overall system. Of course, in order for the application to be available, it must not only have processing resources, but it must also have access to the application data. Therefore, in order to ensure application availability, there must be a copy of the application database at at least two of the nodes; and all processing nodes must have access to all copies of the database, as shown in Figure 3-1.

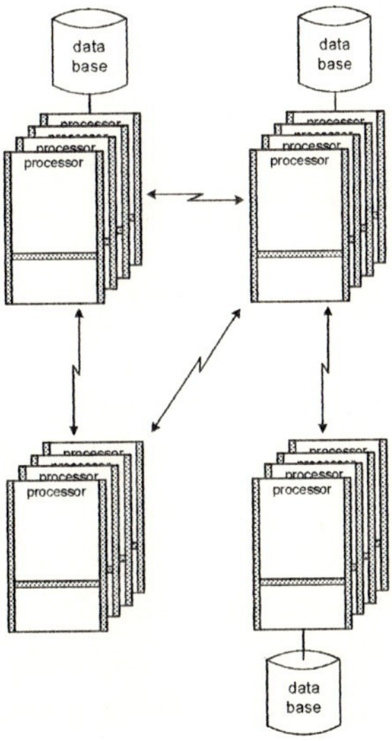

Not all nodes need to have a local copy of the database, and not all nodes need to be directly connected to each other. What is required is that each node have a path to all copies of the database, wherever they may be, so that the results of each transaction can be reflected in all database copies. That is, the various copies of the databases need to be kept in the same state so that the use of any database copy by any node will provide the same result.

**Distributed Application
Figure 3-1**

Therefore, any change made to one database copy must be immediately propagated to all of the other database copies. We call this *data replication*, and its purpose is to keep synchronized the various copies of a database in a redundant system.

There are several ways to do data replication – asynchronously, synchronously, physically, and others. In this chapter, we discuss asynchronous replication, and we discuss synchronous replication in the next chapter.

Uses for Data Replication

There are many advantages to distributing an application across multiple independent nodes:

Disaster Tolerance

The processing nodes may be geographically distributed. In this case, should one node be disabled or destroyed by a natural or man-made disaster, the application will survive and continue to function with no data loss, though perhaps in a degraded mode.

Increased Availability

Even if the nodes are not geographically distributed, the overall system availability can be significantly enhanced by breaking up the system into multiple smaller nodes with no increase in processing or storage resources. This is directly due to a reduction in the number of failure modes for the system. For instance, as shown in Chapter 2, if a large monolithic system is separated into k independent nodes, then the total system availability is increased by at least a factor of k. Furthermore, should there be a failure of a node, only $1/k$ of the system capacity is lost. The chance of losing more than $1/k$ of the total system capacity is virtually never.

Localization

If the users of an application are widely distributed geographically, then processing and data storage resources can be provided much closer to each user, thus improving application response time. In fact, each user community can be served by a node whose size and configuration are more closely attuned to the needs of that user community. In addition, the vendor hardware, operating system, and database product at each node can be chosen independently to minimize costs, to improve manageability, or to otherwise optimize the node for its particular user community.

System Maintenance

If a monolithic system needs to be updated or otherwise maintained, it often must be brought down, thus contributing to undesirable scheduled down time. If the system is distributed, then system maintenance can be applied to one node at a time. During this activity, only one node in the system needs to be down. All other nodes continue to provide full service to the user community.

Enterprise Application Integration (EAI)

Many enterprises are now integrating their disparate applications running on systems from different vendors into a common network for increased functionality, efficiency, and customer service. Many use middleware products to do this. However, this approach generally means that the applications have to be modified to interface properly with the middleware facility. Heterogeneous data replication can provide up-to-date copies of commonly used data on disparate databases in the format expected by each database with no application modifications required.

Operational Data Store (ODS)

One powerful technique for EAI is to implement an independent operational data store that is central to all systems in the enterprise. Any data with immediate value to the enterprise is replicated instantly to the ODS. Data in the ODS is available to all applications no matter where they may reside, thus giving these applications a current view of all data within the enterprise.

Data Warehousing

In a manner similar to supporting an ODS, data replication can be used to feed a data warehouse. Many data replication engines allow the data to be extensively manipulated before applying it to the target database, a requirement for data warehouses that typically archive summary results only.

Dr. Bill Highleyman, Paul J. Holenstein, and Dr. Bruce Holenstein

Zero Down Time System Migration

Data replication can be a powerful way to migrate an application from one system or database to another with zero down time. In principal, the new system is synchronized with the old system while users continue normal activity on the old system. When the new system is synchronized, the users may be switched to the new system. If desired, the new system may continue to replicate its changes to the old system so that a return can be made to that system should the new system experience problems.

Types of Data Replication

In general, a replicated system will have multiple nodes as shown in Figure 3-1, and data replication will be ongoing between all nodes that contain part or all of the database. However, each of these replication paths is commonly served by an independent data replication engine. It is the concurrent operation of all of these replication engines that maintains synchronism between the databases that are distributed across the application network.

Therefore, it suffices to talk only about a single replication engine. A distributed application comprises two or more processing nodes using one or more replication engines to keep its database copies synchronized.

There are many ways to implement a data replication engine, and each has its strengths and weaknesses. We consider in this section five important replication engine characteristics:

- directionality
- threading
- queuing
- target updating
- synchronism

Directionality

By directionality, we mean whether data replication between two nodes is done in only one direction (unidirectional) or in both directions (bi-directional active/active); and if bi-directional, can a particular data item be updated by only one node (partitioned active/active) or by any node (non-partitioned active/active). These various cases are shown in Figure 3-2:[19]

- **Unidirectional**: Unidirectional data replication is typically provided for disaster recovery, data warehousing, and operational data stores. These configurations generally comprise a primary node that is providing all processing functions and a backup node that is ready to take over these functions should the primary node fail, as shown in Figure 3-2a. A current copy of the database is maintained at the backup node by replicating all changes made to the database at the primary node to the backup node. For fast recovery, the backup node may also be configured for replication to the primary node; but this path is inactive unless the backup takes over.

- **Bi-Directional Partitioned Active/Active**: A more efficient use of a replicated system, known as *active/active* or *cooperative processing*, is for all nodes to be cooperating in the processing of the application's transactions. All nodes replicate their changes to all database copies at other nodes. One problem with cooperative processing occurs if two users at different nodes attempt to modify the same data item at the

---

[19] Standish Group calls these architectures Functional Segmentation (Unidirectional Replication), Application/Database Segmentation (Bi-Directional Partitioned Active/Active Replication), and Live/Live (Bi-Directional Non-Partitioned Active/Active Replication). In a poll of high-availability users, Standish Group found that 81% favored the active/active approach. See Standish Group Research Note, "The New High Availability – A NonStop Continuous Processing Architecture (CPA);" 2002.

same time. This problem is analyzed later, but it is avoided if the database can be partitioned so that each data item is in effect *owned* by one and only one node. Only the owner of a data item can update that data item, thus avoiding conflicts. This is the case of partitioned data replication, as shown in Figure 3-2b.

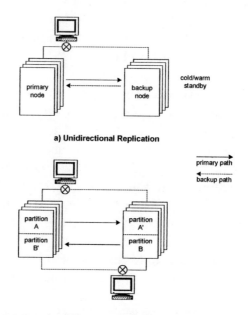

a) Unidirectional Replication

b) Bi-Directional Partitioned Active/Active Replication

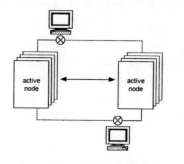

c) Bi-Directional Non-Partitioned Active/Active Replication

Replication Directionality
Figure 3-2

- **Bi-Directional Non-Partitioned Active/Active**: The most general case for cooperative processing is to allow any user at any node to modify any data item in the network, as shown in Figure 3-2c. This is the most flexible and powerful case of cooperative processing, and it is known as active/active data replication. However, in this case, data access conflicts may occur, sometimes with alarming frequency. Means must be provided to resolve data conflicts. Some are described later in this chapter (for data collisions) and others in Chapter 4, "*Synchronous Replication*," (for deadlocks). The potential frequency for data conflicts is analyzed in some detail in Chapter 9, "*Data Conflict Rates*."

Threading

In some applications, the peak rates for data replication may overwhelm a simple data replication engine. In this case, multiple replication threads may be used, as shown in Figure 3-3.

Each thread represents an independent flow of database modifications from the source database to the target database, thus substantially increasing the capacity of the database engine. However, since modification event flow over each thread is independent of all other threads (absent any inter-thread coordination), there is no guarantee that any inherent order in these events will be maintained at the target node.

It may be (and usually is) imperative that some means be provided to ensure a certain order in the applying of these modifications to the target database in order to prevent data corruption or inconsistency. This problem is discussed briefly later in this chapter and is extensively analyzed in Chapter 10, "*Referential Integrity*."

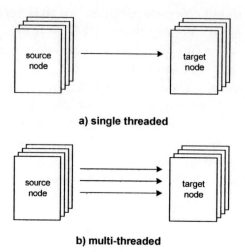

**Data Replication Threads
Figure 3-3**

Queuing

There are many implementations of replication engines. Some are designed to be very fast in terms of getting source database changes applied to the target database. Others queue changes at various points throughout the replication engine for a variety of reasons, as shown in Figure 3-4:

- to minimize the amount of lost data in the event of a network or node failure.

- to provide for proper serialization of changes to the target database.

- to control the applying of changes that are provided by non-recoverable triggers at the source node.

- to prevent transactions that are ultimately aborted from being applied to the target database.

- to improve efficiency of certain components (e.g., blocking messages over a communication channel).

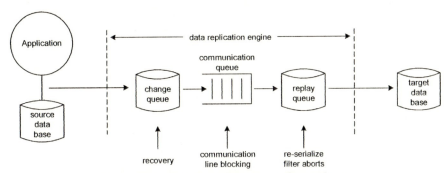

Data Replication Engine Queuing
Figure 3-4

One form of queuing is typically always required, and that is to queue source database changes on a persistent store such as disk at the source node so that they can be replayed following a network or target node failure. True, the persistent store does not need to be used during normal operation; but it must be put into play in the event of a network or target node failure in order to save database changes for later replication.

A replication engine with no queuing points represents the fastest architecture. To the extent that there are one or more queuing points, the interval between the time that a change is applied to the source database and the time that it is applied to the target database is lengthened. This interval is called the *replication latency*. Replication latency affects data replication in many ways and is described later in this chapter as well as in other chapters.

Target Updating

There are several algorithms for how changes are applied to the target database. Each has its own merits in certain applications.

- **Absolute Replication:** When using absolute replication, the data replication engine sends the after images to the target database; and these are used to overwrite the current data items. After images may represent an entire row or record or just the changed data within a row or record. (For deletes, only the row or record key needs to be sent.)

- **Relative Replication:** Alternatively, only the relative differences or change operations are replicated. These differences may be determined by comparing the before and after images of the changed data at either the source node or target node. The relative difference is, in effect, an operation on the target data. For instance, if a source data field is changed from 10 to 15, the replicated relative value will be "add 5" rather than "replace with 15" as is the case for absolute replication.

- **Fuzzy Replication:** Fuzzy replication can be advantageous if the databases are not kept in exact synchronization. If certain replication operations fail, then they are mapped to other operations and retried. For example:

    - if an insert fails because the record already exists, an update is done instead.

    - if an update fails because the record does not exist, an insert is done.

    - if a delete fails because the record does not exist, the operation is ignored; or perhaps a row or record is reinserted with the relative diffcrence.

Synchronism

We have characterized data replication as a means to keep distributed databases synchronized. A more accurate characterization might be to say that data replication will keep database copies exactly synchronized or nearly synchronized. The degree of synchronization depends upon the characteristics of the replication engine.

Specifically, database replication techniques include synchronous replication, that will keep the database copies exactly synchronized,

Breaking the Availability Barrier

and asynchronous replication, that will keep the database copies nearly synchronized. Each has its benefits and problems.

- **Synchronous Replication** is shown simply in Figure 3-5a. In the simplest of terms, all changes to all copies of the database are treated as a single transaction. Either all are made, or none are made. More specifically, the replication engine will obtain locks on all data items in all database copies. It will then make all changes before releasing the locks (i.e., the transaction is committed). If it cannot make all changes, it will restore the original values of the data items before it releases its locks (i.e., the transaction is aborted). In this way, the view of any database copy at any point in time is the same as the view of any other copy. The databases are kept in exact synchronization. Gray[20] calls this *eager replication*.

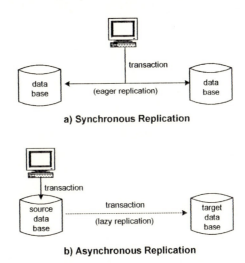

Replication Synchronism
Figure 3-5

Clearly, by behaving in this manner, the application is

---

[20] Gray, J. et al.; *"The Dangers of Replication and a Solution,"* <u>ACM SIGMOD Record</u> (Proceedings of the 1996 ACM SIGMOD International Conference on Management of Data), Volume 25, Issue 2; June, 1996.

somewhat delayed because it must coordinate the modification of two or more database copies across the network. The additional time that it takes a transaction to complete in this distributed environment as compared to the time that it takes in a system with a single database copy is called *application latency*.

- **Asynchronous Replication** is shown in Figure 3-5b. In this case, the application is not held up by the data replication process. Rather, the application operates strictly on a single database instance. Behind the scenes, the data replication engine is monitoring changes to the source database and is sending these changes to be applied to the target database(s). The application is totally unaware of this activity.

  In the above case, there is no application latency. However, there is a time delay from the time that a change is made to the source database and the time that the change is applied to the target database. This time delay is known as *replication latency* and was described earlier with respect to replication engine queuing. Thus, the target database lags the source database by the replication latency time; and the database copies are *nearly* synchronized. Gray refers to this as *lazy replication*.

The rest of this chapter deals with issues relating to asynchronous replication. Synchronous replication and application latency are discussed in Chapter 4, "*Synchronous Replication*."

The Basic Data Replication Engine

Before discussing some of the issues with respect to asynchronous replication, let us look at the basic architecture of a data replication engine. Architectures for replication engines are described in much more detail in Chapter 10, "*Referential Integrity*."

A replication engine generally has three main components, as shown in Figure 3-6a:

Breaking the Availability Barrier

- an Extractor on the source node. The Extractor monitors the source database for changes.

- a communication channel that the Extractor uses to send changes to the target node.

- an Applier on the target node. The Applier receives changes from the Extractor and applies them to the target database.

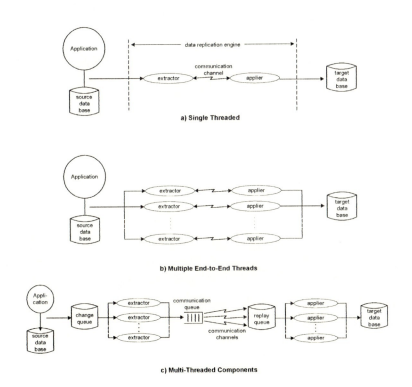

Data Replication Engine Architectures
Figure 3-6

Although Figure 3-6a shows the Extractor on the source node and the Applier on the target node, this is unnecessary. The Applier can also be on the source node (and perhaps be in the same process as the Extractor) and will "push" database changes to the target database via remote procedure calls (RPCs) or some similar mechanism. Likewise, the Extractor can reside on the target node, can be part of the Applier, and will "pull" database changes over the communication channel.

Figure 3-6a illustrates a single-threaded data replication engine. Alternatively, for performance purposes, the replication engine can be multi-threaded. Figure 3-6b shows multiple end-to-end threads that can operate independently to carry their own stream of database changes. Not shown is a mechanism for re-serializing these changes into proper order if needed before applying them to the target database.

Figure 3-6c shows a more complete picture of a different type of multithreading. It shows that each of the replication engine components – the Extractors, communication channels, and Appliers – can be individually multithreaded. Any combination of component single-threading and multithreading can be provided.

In addition, major optional queuing points are shown in Figure 3-6c. They are only implied in Figures 3-6a and 3-6b:

- **Change Queue:** Changes extracted from the source database are commonly written to a persistent store such as disk so that they can be replayed in the event of a network or target node failure. During peak transaction rates, this queue also helps buffer changes that might otherwise overwhelm the replication engine. The Change Queue can be found in many forms, including:

    - an Audit Trail maintained by the node's transaction manager.

    - a Change Log created by the application.

    - a Database of Change (DOC) file created by the replication engine if there is no external Change Queue created outside of the replication engine.

- **Communication Queue:** Changes are often held at the communication channel and accumulated so that large blocks of changes may be sent over the communication channel. In this way, the replication engine sends an occasional (relatively speaking) large message rather than many frequent small messages. This will, in many cases, significantly increase the efficiency of use of the communication channel and will reduce communication delays.

- **Replay Queue:** If multithreading is used, re-serialization of changes arriving at the target node may be required to ensure that changes are applied to the target database in the correct order. One way to achieve this is by storing all changes as they arrive in a Replay Queue (that may be disk-resident). Changes are then read them from the Replay Queue in correct order using one or more closely coordinated Appliers, that will apply the changes to the target database node in the correct order.

The Replay Queue or the Change Queue can also be used to store an entire transaction before applying it or sending it to the target node. In this way, changes that are part of transactions that are later aborted need not be applied or sent to the target database. This is especially useful if the target database does not support the notion of transactions, that means that a partially applied set of changes cannot be backed out or aborted.

Advantages of Asynchronous Replication

There are several important and positive characteristics that are typical of these asynchronous replication architectures.

No Performance Penalty

The addition of an asynchronous replication engine to an application generally imposes little application performance penalty since all replication activity is independent of the application. Replication is totally decoupled from the applications and interacts with them only via the Change Queue.

Non-Invasive

The addition of a data replication engine to an existing application is often non-invasive. It requires no changes to the application. The data replication engine acquires source database changes through mechanisms that often already exist (such as an Audit Trail). Therefore, it is easy to retrofit existing applications for disaster recovery or for EAI or to distribute them to improve availability or locality.

Heterogeneous

Replication may be between heterogeneous databases or even between heterogeneous systems since the Extractors and Appliers are loosely coupled through a messaging system (the communication channel) and can be tailored to their own environments. For instance, the source node can be a legacy system using flat files; and the target node can be an open system using a relational database.

Data Manipulation

The replication engine can provide virtually any data manipulation function of the data prior to applying it to the target database. This can include simple functions such as data format conversion or complex functions such as table look-up, aggregation, and computation. Data manipulation is application dependent. It may be supported by scripts or user exits usually made available within the source or target node replication engine components such as the Extractor or the Applier.

Data Integrity

The replication engine will ideally apply each event once and only once to the target database. The replication engine can be configured so that the applying of modifications and transactions is in the same order as that at the source database. The preservation of the *natural flow* of database modification events guarantees that the target database will always be an exact copy of the source database, though perhaps somewhat delayed by the replication latency.

Highly Reliable

The replication engine can be designed to be highly reliable in that a source node will never lose any changes as long as they are stored in a persistent store. In this way, the full recovery of a downed or isolated node now restored to service can always be guaranteed by replaying changes that it had not received.

Highly Available

The replication engine can be designed to be highly available, especially if running in a fault-tolerant environment. Extractors and Appliers can be implemented as primary/backup process pairs running in different processors. Alternatively, they can be configured as persistent processes that will be restarted (in a different processor if need be) following a failure. This function is usually provided by a fault-tolerant monitor process pair or as an operating system service.

Highly Scalable

The replication engine is inherently highly scalable. Since each replication engine handles just one source/target database instance, adding database instances to the application network does not increase the load on any one replication engine. Rather, additional replication engines are simply added for each additional node pair. For instance, a two-node system will have two replication engines, one for each direction. A four-node system will have at most twelve replication engines, one pair for each of the six possible connections.

Highly Secure

The replication engine is decoupled from the source applications and the source database and so does not provide a path into the source node for unauthorized access. Furthermore, the specific data to be shared with other nodes can be tightly controlled.

Asynchronous Replication Issues

Discussed above are some of the many advantages of asynchronous replication. However, there are also many issues with that to be concerned. All can become non-issues if handled correctly.

General Issues

Some of these issues are independent of the type of asynchronous replication. They include the consequences of replication latency and multithreading.

Data Loss

The fact that there is replication latency associated with asynchronous replication means that at any one time there are transactions in the replication pipeline flowing from the source node to the target node. Should a node fail, the transactions that it had in the pipeline are lost. True, they can be recovered from the source node's Change Queue once that node has recovered. While the node is down, however, the transactions will be unavailable.

The amount of data that may be lost is directly proportional to replication latency. Minimizing replication latency will minimize the potential for data loss in the event of a source node failure.

Rule 11: *Minimize data replication latency to minimize data loss following a node failure.*

However, if this temporary loss of transactions is considered a serious and perhaps unrecoverable problem, then synchronous replication as described in the next chapter is a good solution. Synchronous replication guarantees that no data is ever lost.

Database Corruption

It is often very important to apply changes to the target database in the same order as they were applied to the source database. This is called the *natural flow* of changes and transactions. If natural flow is violated, there are several opportunities for database corruption.

For instance, if two absolute updates to the same row or record are applied in reverse order, the data item will be left at the older value. Or a child record may be inserted before a parent record, causing a referential integrity violation. This may cause the target database to reject the child insert.

> **Rule 12:** *Database changes generally must be applied to the target database in natural flow order to prevent database corruption.*

A single-threaded replication engine as shown in Figure 3-6a will typically guarantee that the event order delivered to the target database is identical to the event order received from the source database. However, if the replication engine is multithreaded, as shown in Figures 3-6b and 3-6c, there is no guarantee of order at the target database since changes and transactions may flow over the different threads at different rates. Therefore, a mechanism for re-serialization must be provided following all multithreaded activity and prior to the target database. This is exemplified by the Replay Queue shown in Figure 3-6c.

Interestingly, multithreading is employed to increase the capacity of a data replication engine. However, even though the capacity of the replication channel may increase when multithreading is used, the requirement for re-serialization may increase the actual replication latency depending upon the re-serialization method employed.

One technique to avoid database corruption due to improper modification event order is to hold up the replication of a transaction until all previous transactions have been committed. However, this might not only slow down the replication channel but also may cause peak loads that were not observed on the source database to occur on the target database.

> **Rule 13:** *Follow natural flow order when replicating so as not to create artificial activity peaks at the target database.*

The effects of multithreading and techniques for ensuring that multithreading does not cause data corruption are extensively analyzed in Chapter 10, *"Referential Integrity."*

Ping-Ponging

Ping-ponging is an issue when bi-directional replication is used. If care is not taken, a change replicated from a source node to a target node can be replicated back to the source node, then back to the target node, and so on ad infinitum. This effect is also called *oscillation* or *data looping*.

To understand why ping-ponging can occur, consider a replication engine that is being driven by an Audit Trail created by a transaction manager. Whenever a change is made to the database, the transaction manager will enter a description of that change into the Audit Trail. It doesn't care where the change came from. The data replication engine, unless told otherwise, will dutifully replicate that change to its target node, even if it just came from the target node.

> **Rule 14:** *Block the ping-ponging of data changes in a bi-directional replication environment to prevent database corruption.*

In order to prevent ping-ponging, the replication engine must be able to determine the source of the change and replicate only those changes that were made by the local applications. Alternatively, all changes can be replicated back to the source node and ping-ponged

Breaking the Availability Barrier

changes filtered out at the source node. There are several ways that this might be implemented[21], including the following:

- **Partitioning:** If the database can be partitioned so that each data item is owned by one and only one node, and if only the owner can update its own data items, then the replication engine can be implemented to replicate only changes to those data items owned by its node.

- **Data Content:** In many cases, the row or record may indicate the source of the change. This might be, for instance, a source node id. In such a case, the data replication engine will not replicate changes originating at other nodes.

- **Control Table:** If there are no other means for the replication engine to determine the origination point of a change, then it can track remotely originated transactions via a Control Table that it maintains, as shown in Figure 3-7. As an Applier receives transactions and applies them to the target database, it enters a transaction id for that transaction into the Control Table. Likewise, as an Extractor at that node reads transactions from the Change Queue, it will check each transaction to see if that transaction's id is contained in the Control Table. If the Extractor finds the id, then it will not replicate that transaction.

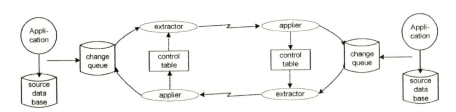

Ping-Pong Avoidance
Figure 3-7

---

[21] Strickler, G.; et al.; *"Bi-directional Database Replication Scheme for Controlling Ping-Ponging,"* United States Patent 6,122,630; Sept. 19, 2000.

Dr. Bill Highleyman, Paul J. Holenstein, and Dr. Bruce Holenstein

Data Collisions

Because asynchronous replication modifies the target database some time after the source database has been modified, there is a time interval during that the data item that was modified is not accurately represented in the target database. This time is called the *stale time* and is predominantly the replication latency time.

Should an application modify a data item while that data item is stale, a *data collision* occurs. The source node is in the process of sending its latest change to the target node while the target node, acting as a source node, is replicating its latest change. Both changes are different and probably are both wrong. The database at this point has been corrupted and must be fixed.

Clearly, the frequency of data collisions will increase proportionally with the stale time. The stale time is predominantly the replication latency time of the replication engine as described in more detail in Chapter 9, "*Data Conflict Rates*." Therefore,

> **Rule 15:** *Minimize replication latency to minimize data collisions.*

There are two ways to solve the data collision problem. One way is to avoid it. The other is to detect data collisions and resolve them.

Collision Avoidance

There are configurations that are not susceptible to data collisions. Among them are:

Partitioned Database

As suggested above as a cure for ping-ponging, partitioning the database will eliminate collisions. To review, each partition in a partitioned database is owned by one and only one node. Only the owner of a partition may modify data in that partition. Since all application changes to a data item are made by only one node, there are no data collisions.

Partitioning may take many forms. It may be based on the value of a particular field. Perhaps only local sales offices can make changes to orders originating in their territory. A cell phone system may route messages to different nodes based on the first digit of the called or calling number. Partitioning might be based on customer and so on.

Partitioning might also be time-based. In a "pass the book" strategy, ownership of a database might pass from one node to another based on some criteria. An international brokerage firm might have the ownership of its database follow the sun, passing the database from office to office as the 24-hour day progresses. Alternatively, the ownership might be rotated between nodes on a schedule so that each node can process change requests that it has batched.

In any event, partitioning ensures that there is only one owner of a data item at any point in time. Therefore, there can be no data collisions.

Synchronous Replication

If partitioning is not feasible, and if data collisions must be avoided, then synchronous replication can be used. It is described in more detail in the next chapter. Synchronous replication ensures that all copies of a data item across the network are locked before any changes are made. Then all changes are simultaneously made before the data items are unlocked. In this way, changes at the target nodes are changed in synchronism with the changes at the source node; and no data collisions can occur.

Note that with synchronous replication, an application that is trying to modify a data item that is being changed by another application anywhere in the network must wait for the lock on that data item to be released before it can make its change. If asynchronous replication is being used, this application can make its change while the other application is making its change, and a data collision will result instead. As Jim Gray noted,[20]

> **Rule 16:** (Gray's Law) *Waits under synchronous replication become data collisions under asynchronous replication.*

Collision Detection

If data collisions cannot be avoided, then one must either decide that they are unimportant and will have no adverse effect or that they must be detected and resolved.

There are several existing techniques for detecting a collision. They are generally based on ensuring that the row or record to be changed on the target database is the same version that was changed on the source database:

Before-Image Comparison

If the replication engine sends before images as well as after images, then the Applier can compare the current state of the row or record on the target node to the before image of the record or row that was modified at the source. If they are the same, then a collision has not occurred; and the after image may be applied to the database.

Versioning

A single field indicating the original version of the source database record or row that is being modified can be sent with the change data. The Applier can then verify that the target record or row is the same version. There are many ways to check versions, such as a date-time stamp, an embedded version number field or column, or the CRC code associated with the row or column.

Collision Resolution

The resolution of data collisions is highly application-dependent, and the rules for resolution may be different for different fields. Each application must be carefully analyzed, and the business rules for resolving the results of a data collision for each field or column in each file or table must be determined. These rules may include the following.

Generic Algorithms

Many data collisions can be resolved by broad generic algorithms. For instance:

- always use the change from a designated master node.

- always use the change with the most recent or least recent time stamp or with the highest or lowest version number.

Data Field Specific Algorithms

One may be able to establish resolution algorithms that are specific to certain data types and/or data fields. For instance:

- for date/time fields, always use the earliest or latest date/time.

- for numeric fields, use the minimum, maximum, average, or some other function.

- for text fields, application-specific rules may be able to be formulated.

- for fields that have no real application significance, ignore the collision.

- for fields that may have no local significance, ignore the collision.

Relative Replication

For many numerical fields, such as counters, accumulators, and dollar quantities, relative replication may be used. The change in the numerical value can be computed by comparing the before and after images, and then that change can be applied to the field.

Care must be taken when applying relative replication. Only *commutative* operations may be replicated. Commutative operations are those that can be applied in any order and can still achieve the correct result. Addition and subtraction are commutative and may be replicated as differences, as described above. Multiplication and division are commutative and may be replicated as factors. For instance, if node A divides a data item by 5, and if node B multiplies it by 2, then node A can replicate a ÷5; and node B can replicate a x2.

However, addition/subtraction and multiplication/division are not commutative (10+2x5 is not equal to 10x5+2). Furthermore, if a field is subject to both addition and multiplication, one cannot determine from the before/after images whether the field was modified by addition or by multiplication. Therefore, the replication engine will not know whether to replicate differences or factors.

One problem with relative replication has to do with aborts. If commutative operations are applied relatively, then the operation can be saved as an effective before image. If the transaction is aborted, then the operation is simply reversed – i.e., an add becomes a subtract, a multiply becomes a divide, and so on. These are called *symmetric* operations. However, there are many operations that are *asymmetric* and cannot be easily backed out. The selection of the latest date or the maximum value of a numeric field are examples of asymmetric operations.

Fuzzy Replication

Fuzzy replication is the term given to a replication mode that re-performs a failed operation by mapping it to another operation in a defined way. For instance:

- If the operation is an insert that has found that the data record already exists, then it can be converted to an update using a mix of the above rules.

- If the operation is an update, and if the record does not exist, then it can be ignored (give precedence to the presumed delete with that it collided). It also can be converted into an insert, depending upon business rules.

- If the operation is a delete, and if the record does not exist, then it can be ignored. Alternatively, it may be converted into an insert with the relative difference applied.

Note that some of these sequences may repeat due to additional collisions, though with less and less likelihood as the sequence goes on. For instance, a collision that causes a data item to be deleted and then reinserted may continue if the deletion is retried in the face of another update.

Manual Resolution

If all else fails, the data collision must be sent to a human operator for manual resolution. In this case, the result will be very much dictated by business practices.

Failures and Recovery

Failover

One of the primary purposes of data replication is to provide continuing service to the users even in the presence of a node or network fault. What makes this possible is that following a failure, there is always at least one node still functioning in the network; and that node has access to a current copy of the database. The downed applications can move to surviving nodes. We call this procedure *failover*. Failover procedures vary somewhat according to the type of failure. Reference should be made to Figure 3-2.

Source Node Failure

- **Unidirectional Replication:** In the event of a source node failure, the backup node has a reasonable up-to-date copy of the database, albeit with the possibility of uncompleted transactions, that must be aborted. This is because the backup node may not have received certain transactions or their commits, that were still in the replication pipeline at the time of the source node failure. Normal processing may be resumed as soon as the users have been switched from the primary node to the backup node and as soon as the incomplete transactions have been aborted (to release locks that those transactions are holding). This switchover need take only a few seconds if the communication infrastructure is in place. Once switched over, users should each check that their last transaction has been applied to the database. Transactions lost in the pipeline should be resubmitted.

Breaking the Availability Barrier

- **Bi-Directional Replication:** Whether bi-directional replication is being handled via partitioning or is active/active, only the users on the downed node are affected. All users on the surviving nodes are unaffected. As soon as the downed users are switched to a surviving node and check their last transactions, normal operations continue. If the increased load on the surviving nodes is a problem, load shedding by terminating non-critical applications may be in order.

Target Node Failure

- **Unidirectional Replication:** If the backup node fails, there is no impact on users. The users, who are all connected to the primary node, continue to be provided with full application services.

- **Bi-Directional Replication:** Users at the surviving nodes are unaffected. Failover for the users at the failed node is identical to that described above for a source node failure.

Network Failure

Should the communication channel fail between a pair of nodes, the nodes are left operational; but replication will cease. During this time, the nodes will queue their database changes to their Change Logs for use during later restoration.

- **Unidirectional Replication:** In a unidirectional replication environment, there is no further action to take. Users remain active on the primary node.

- **Bi-Directional Replication:** There are many strategies that may apply to failover following a network failure in a bi-directional environment. One strategy is to allow users at isolated nodes to continue their normal functions. The replicating engines will queue all database changes to the isolated node for later restoration.

- However, during the network down time, if the network is not partitioned, there is the possibility that many data collisions will occur and will need to be detected and resolved at restoration time. If collisions are a problem, then isolated nodes should be considered to be failed nodes; and users should be switched to other nodes as described above.

Restoration

Restoration of full services following the return to service of a failed or isolated component is accomplished by resynchronizing all database copies in the network. During the failure period, all active nodes were queuing their database changes for later replication. Now, they will each begin replaying these changes to the previously downed or isolated nodes (that can include two-way replication if two or more nodes were isolated by a network failure). When all change queues have been drained to an acceptable level from a replication latency viewpoint, and when the users have been switched back to their home nodes, normal system operations resume.

In order to replay changed data, the source replication engine's Extractor must know where successful replication left off. It does this by keeping track of the Change Log's oldest transaction still outstanding at the time of the failure (to support this, an Applier might return a confirmation to the Extractor when it has successfully applied a transaction). Periodically, the Extractor may checkpoint this restart point to disk.

When a replay is required, the replication engine will begin sending transactions starting at the replay point. Normal operations can continue at that node since subsequent database changes will simply be added to the tail of the Change Queue. When the Change Queue has been drained to the point that replication latency has returned to a normal level, synchronization of the database copy has been completed.

There are, however, some critical issues associated with replay:

Duplicate Transactions

It is quite possible that transactions that have already been applied to the target database will be replayed again during restoration. There are two reasons for this. One is due to the fact that transactions are interleaved. Starting replay at the oldest open transaction implies that there are more recent transactions that have completed and that will be replayed. Also, if a disk checkpointed restart point is used, there may have been significant transaction activity following the checkpointed replay point; and that activity will all be re-replicated.

In most cases, transactions cannot be duplicated. Most are not *idempotent* – that is, multiple applications of the same transaction will yield different results. To correct this situation, duplicate transactions must be detected and ignored. One way to do this is for each Applier to maintain a persistent record at the target node of all transaction ids that have been successfully applied to the target database. By including updates to this table within the scope of the target transaction, a transaction id will be entered only if the transaction successfully committed.

On replay, the Applier will check each incoming transaction against this table to see if the transaction is a duplicate and will discard it if it is.

Data Collisions

If replication is unidirectional or if it is bi-directional partitioned, then there is no problem with data collisions while the downed database is being resynchronized. However, if replication is active/active, and if a network failure isolated one or more nodes that remained active during the failure (i.e., users were not switched over to connected nodes), then data collisions may occur during the network down time. In effect, the stale time has been extended from the normal replication latency time to the network failure time.

These data collisions are no different than normal data collisions that occur during normal asynchronous replication stale time, except that a lot more have been accumulated. They must be detected and resolved during replay, as described earlier.

What's Next?

In this chapter, we have discussed various aspects of asynchronous replication. Two characteristics of asynchronous replication are that all copies of the database are not kept in exact synchronization – they may differ by the replication latency time - and as a result, data collisions may occur. The copies may also suffer data loss in the event of a failure.

If exact synchronization is called for, or if data collisions or data loss are intolerable, then synchronous replication is a serious consideration. However, synchronous replication brings with it a potential performance impact in that an application must wait for a transaction's changes to be sent and applied to all database copies across the network before it can commit the transaction. This additional application delay time is known as *application latency*.

Various methods for synchronous replication are described in Chapter 4, "*Synchronous Replication*," and the application latencies associated with these methods are analyzed and compared.

Chapter 4 - Synchronous Replication

In Chapter 2, "*System Splitting*", we looked at the availability advantages of replicating or splitting a processing system. If a system is split into multiple independent nodes, it is important to have two or more copies of the database distributed across the network in order that application data is available to surviving nodes in the event of a node failure.

These database copies must all be kept in synchronism so that all applications running on all nodes will have the same view of the application data. This synchronism can be easily achieved via asynchronous data replication – there are many data replication products available today that provide this capability without having to modify the applications. However, if all nodes are actively participating in the processing of transactions, data collisions may occur at an uncomfortable rate due to replication latency. These data collisions, unless detected and resolved, will result in database contamination.

Data collisions can be avoided by synchronously updating all database copies so that all copies of all data items are always guaranteed to be the same. This can be accomplished with synchronous data replication, which is the subject of this chapter.

Replicating Systems

The replication of systems is an important and popular technique to ensure that critical computing systems will survive system failures caused by anything from component failures to man-made or natural disasters.

Dr. Bill Highleyman, Paul J. Holenstein, and Dr. Bruce Holenstein

Under a typical system replication scenario, one system is the primary and handles the entire transaction load. As shown in Figure 4-1a, updates made by the primary system to its database are replicated to the backup system. The backup system is passive except for perhaps supporting read-only operations such as query and reporting.

As shown in Chapter 1, *"The 9s Game,"* fully replicating a system, in addition to providing protection from disasters, has the effect of doubling its 9s, thus dramatically improving the system's availability.

Splitting Systems

Alternatively, significant availability advantages can be achieved by simply splitting a single system into k nodes. In Chapter 2, we showed that doing so could increase system reliability (i.e., increase its mean time before failure, MTBF) by more than a factor of k.

However, when we split a system (for example, splitting a 16-processor system into four 4-processor nodes), all nodes must be actively sharing the load. This implies that all nodes are updating the database.

Often, the database can be partitioned so that only one system can update a given partition; and those updates can be replicated to the other systems for read-access only (Figure 4-1b). In this case, the most serious concern is *replication latency*, or the time that it takes for an update to propagate from the source node to the target node. Updates in the replication pipeline may be lost in the event of a system failure.

However, in the more general case, any system in the network must be able to update any data item (Figure 4-1c). Those updates then must be replicated to the other databases in the network. We call these *active/active* replication applications. In addition to the problems imposed by replication latency, as described above, active/active applications present additional significant problems. One

Breaking the Availability Barrier

of the most severe problems is data collisions. To the extent that there is replication latency, there is a chance that two systems may update the same data item in different copies of the database simultaneously. These conflicting updates then will be replicated across the network and result in a corrupted database (i.e., the value of the data item will be different in different instances of the database).

For instance, an application at node A might change the value of a particular data item from 10 to 15. At nearly the same time, an application at node B might change that same data item from 10 to 20. Node A will then replicate its value of 15 to Node B, which will set the data item value to 15. Likewise, Node B will replicate its value of 20 to node A, which will set the value of the data item at its node to 20. Now the data item not only has different values at the two nodes, but both values are wrong.

We call these simultaneous conflicting updates *data collisions*. Data collisions must be detected and resolved, often manually.

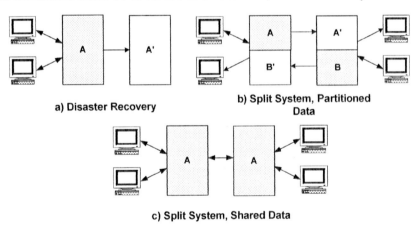

**Split System Architectures
Figure 4-1**

Dr. Bill Highleyman, Paul J. Holenstein, and Dr. Bruce Holenstein

Data Collisions

The probability that data collisions will occur is surprisingly high. An extension of Jim Gray's work[22], presented later in Chapter 9, "*Data Conflict Rates*," shows that the data collision rate in such a network is given by

$$\text{Data Collision Rate} = \left(\frac{d-1}{d}\right)\frac{(kra)^2}{D}(L+t/a)$$

where

- r is the transaction rate generated by one node in the network,
- a is the number of replicable actions in a transaction (updates, inserts, deletes, and in some systems read/locks),
- t is the average transaction time,
- k is the number of nodes in the network,
- L is the replication latency time,
- d is the number of database copies in the network,
- D is the size of the database (in terms of data objects – i.e., the lockable entities).

Consider a system split into four nodes with a requirement to maintain an up-to-date database at each node ($k=d=4$). Let us assume that each node generates a leisurely ten transactions per second ($r=10$) and that an average transaction involves four updates ($a=4$) and requires 200 milliseconds to complete ($t=.2$). Furthermore, let us assume that our database requires 10 gigabytes and that an average row (the lockable entity) takes 1,000 bytes. Thus, the database contains 10 million lockable objects ($D=10,000,000$). Finally, the replication latency is 300 milliseconds ($L=.3$).

---

[22] Gray, J.; et al.; "*The Dangers of Replication and a Solution*," ACM SIGMOD Record (Proceedings of the 1996 ACM SIGMOD International Conference on Management of Data), Volume 25, Issue 2; June, 1996.

Using the above relation, we find that our system will create over 2.4 collisions per hour. This can be a major headache. If the nodal transaction rate increases to 100 transactions per second, the collision rate jumps to over 240 collisions per hour. This will certainly keep a team of people busy. If the system then grows to eight nodes, the collision rate will explode to over 1,100 collisions per hour. This is untenable.

Data collisions first must be detected and then must be corrected either manually or by automatic conflict resolution via business rules. Collision detection methods not only add overhead to the replication engine, but they also are only the start to collision resolution. The correction and resynchronization of the database is often a lengthy and manual operation since automated resolution rules are often not practical.

Synchronous Replication

In this chapter, we deal with methods for avoiding data collisions in active/active applications rather than having to correct them.[23] By implementing collision avoidance, there is no need for a collision detection mechanism; nor is there need for a resolution strategy that may involve complex business rules.[24]

The avoidance of collisions requires that all data items be updated synchronously. That is, when one copy of a data item is updated, no other copies of that data item can be changed by another update until all other copies have been similarly updated as well. We call this *synchronous replication*.

Synchronous replication carries with it a performance penalty since the transaction in the originating node may be held up until all

---

[23] Holenstein, B. D.; et al.; "*Collision Avoidance in Data Replication Systems,*" United States Patent Application No. 2002/0133507; Sept. 19, 2002.
[24] This is an oversimplification. As pointed out later, network failures and node failures require resynchronization of the databases in order to recover. However, this statement is valid for normal operations.

data items across the network have been updated. In this chapter, we explore the performance impact of synchronous replication.

We consider two techniques for synchronous replication – dual writes and coordinated commits. To simplify the analysis, we consider a system comprising only two nodes. This analysis is extended later to systems comprising multiple nodes, and the conclusions are the same.

Dual Writes

The term *dual writes* is just another name for network transactions. Dual writes involve the application of updates within a single transaction to all replicates of the database. This is accomplished by using a two-phase commit protocol under the control of a distributed transaction manager. The transaction manager ensures that all data items at all sites are locked and are owned by the transaction before any updates are made to those data items, and it then ensures that either all updates are made (the transaction is committed) or that none of them are made (the transaction is aborted). In this way, it is guaranteed that the same data items in different databases will always have the same value and that the databases therefore will always be consistent.

A simple view of synchronous replication using dual writes under a transaction manager is shown in Figure 4-2. A transaction is started by the application, and updates are made to the source database and also to the target database across the network (1). The updates to the target database may be generated directly by the application, may be generated by an intercept library, or may be generated by database triggers that invoke, for instance, a stored procedure when an update is made. Each updated data item is locked, and the locks are held until the completion of the transaction. When all updates have been completed, the transaction will be committed. At this time, the transaction manager will apply all updates to the source and target databases (2). If the transaction manager is unsuccessful in doing this, then the transaction is aborted; and no updates are made.

Breaking the Availability Barrier

There is a very small window of confusion that may cause the outcome of a transaction to be uncertain under some failure conditions. This is a characteristic of the two-phase commit protocol commonly used by transaction managers. Typically, when the originating application is ready to commit, the transaction manager will ask each node involved in the transaction if it is ready to commit. If all nodes concur that they are holding all locks and have safe-stored all modifications (completion of phase 1), then the transaction monitor will command all nodes to commit (phase 2). Should the network fail after a remote node has responded favorably to the phase 1 "ready to commit?" query but before that remote node has received the phase 2 commit command, then the remote node does not know whether the transaction was committed or aborted. This situation must be resolved either manually or by business rules governing the system behavior in this very unlikely situation.

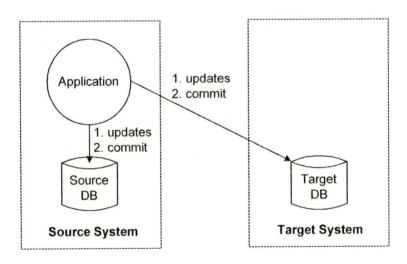

**Dual Writes
Figure 4-2**

Note that transaction updates under a transaction manager may be done either serially or in parallel. If the updates are serial, then the application must wait not only for the local updates to be made, but it must also wait for the update of each remote data item over the network. If updates are done in parallel, then to a first approximation the application is delayed only by the communication channel propagation time. The read/write time at the target database is transparent to the application since it must spend this same amount of time updating the source database.

Most operating systems or databases today do not support dual writes directly. Therefore, to implement dual writes, it is usually necessary to actually modify the application, to bind in an intercept library, or to add triggers to the database to perform the writes for each data item to each database copy. Thus, implementing a dual write solution is invasive to the application in most cases.

Coordinated Commits

An alternative approach to dual writes is to begin independent transactions on each node and then to *coordinate the commits* of those transactions so that either they both commit or that neither commits. In this case, normal data replication techniques (see Chapter 3, "*Asynchronous Replication*") are used to propagate updates asynchronously to the target node. There they are applied directly to the target node's database as part of its transaction that was started on behalf of the source node. In this way, the propagation of updates over the network is transparent to the application.

A simplified view of an implementation for coordinated commits is shown in Figure 4-3. The application first will start a local transaction. As the application locks data items and makes updates to its source database (1), the changes to the database are captured (either by reading an audit file or log file or by intercepting the update commands from the application or the database). These updates are sent to the target node (2), where a transaction is started, locks are acquired, and the updates are made to the target database.

Breaking the Availability Barrier

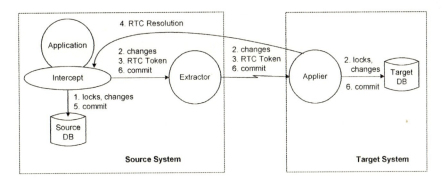

**Coordinated Commits
Figure 4-3**

When the application attempts to commit the transaction, the commit is intercepted (perhaps by an intercept library linked into the application). Before the source node is allowed to commit the transaction, a Ready-To-Commit (RTC) Token is sent to the target node (3) through the replicator to assure that the token will arrive at the target node after the last update. The RTC Token queries whether all of the changes to the target database are ready to be applied. At this point, the target node will respond to the source node with an RTC Resolution message (4). The source node side of the data replicator will release the commit to the source database (5). When this has committed successfully, a commit message is sent to the target node (6), which then will commit the transaction.

Should the target node respond negatively to the RTC Token, then the transaction is aborted at the source node. There is the possibility that the source commit might succeed and that the subsequent target commit fails. Since the target node has guaranteed that it is holding locks on all of the data items to be updated when it returns the RTC Resolution, then this type of failure should occur only if the target node or the network fails after the target node has returned the RTC Resolution and before it receives the commit directive from the source node. This failure window is similar to the window of confusion described earlier for dual writes.

In this case, the normal data resynchronization capabilities of the replication engine will have to be invoked to resynchronize the databases. However, the databases will be out of sync anyway when the target node is returned to service. The lost transaction is just one of many that will be recovered through the resynchronization procedure. Resynchronization facilities are normally provided with data replication engines. However, this is often not the case if dual writes are being used – database resynchronization following a network or target node failure often requires a user-written resynchronization utility.

Note that coordinated commits are implemented with a standard asynchronous data replication engine that has been enhanced to coordinate the commits at the various nodes. Therefore, as with asynchronous replication, an application can be retrofitted to be distributed using coordinated commits without modifying the application. The use of coordinated commits, unlike dual writes, is non-invasive to the application.

In addition, data replication offers the option to enhance the efficiency of data communication channel utilization by buffering the many small messages involved and then sending them to the remote nodes as blocks of changes. When dual writes are used, each database change must be sent individually.

Application Latency

From a performance viewpoint, we are interested in the additional delay imposed upon a transaction due to having to wait for the completion of updates to the target node. We call this additional delay the *application latency* caused by synchronous replication.

Note that if data replication is asynchronous rather than synchronous, then the application will not be delayed by data replication. Instead, the target database will lag behind the source database by a time interval which we have previously called *replication latency*.

Breaking the Availability Barrier

Also note that application latency will increase the response time for a transaction but will not in itself affect thruput. Thruput can be maintained simply by configuring more application processes to handle the transaction load. This solution, of course, assumes the use of well-behaved application models that allow the use of replicated application processes such as NonStop server classes.

Let us now calculate the application latency for dual writes and for coordinated commits so that they can be compared.

Dual Writes

To simplify our analysis of dual write application latency, we assume that all remote database operations are done in parallel with the database operations at the source node. We further assume that all database operations are full updates that require a read/lock of the data item followed by a write rather than simple operations such as read/locks (that is, fetches), inserts, and replacements that require only one access of the database. It is quite straightforward to modify the following relationships to account for these factors. In fact, we do so later in this chapter and will show that the general conclusions reached are unchanged.

In order to update a data item across the network, a read/lock command first must be issued and the data item then received. Next, the updated data item must be returned to the target database and a completion status received. Thus, there are four network transmissions required to complete one update.

In addition, the transaction manager's two-phase commit protocol requires four network transmissions. A prepare-to-commit message is sent and is followed by a response (phase 1). Then the commit message is sent and is followed by its response (phase 2). Only upon receipt of the commit response is the transaction considered to be complete at the source node.

Let

L_{dw} be the dual write application latency,
n_u be the average number of updates in a

transaction,

t_c be the communication channel propagation time, including communication driver and transmission times.

Then

$$L_{dw} = 4n_u t_c + 4t_c \qquad (4\text{-}1)$$

In Equation (4-1), the first term is the communication time required to send the updates to the target node. The second term is the communication time required to commit the transaction.

The above has ignored the processing time required for generating the additional remote database operations and for processing their responses. These times generally are measured in microseconds, whereas channel propagation times typically are measured in milliseconds.[25]

Coordinated Commits

Coordinated commit replication requires the use of a data replication facility to propagate the updates to the target node. Most of the time spent by this facility is invisible to the application providing that updates to the remote database are made as soon as they are received without waiting for the commit (the optimistic strategy).

However, once all updates have been made, the application then must wait for the RTC Token to be exchanged before it can carry out its transaction commit. Note that once the commit has completed at the source node, the subsequent commit at the target node is asynchronous relative to the application. The commit's success is guaranteed since the target node has acquired locks on the data items to be changed, and it will apply the changes upon receipt of a commit directive from the source node. Therefore, the source node does not

---

[25] For a more detailed analysis of processing times, contact the authors.

Breaking the Availability Barrier

have to wait for the target node's commit to complete, and the target node commit thus does not add to application latency.

We estimate the application latency for coordinated commits, L_{cc}, as follows. Let

L_{cc} be the application latency for coordinated commits,
t_p be the processing delay through the replicator exclusive of the communication channel propagation time,
t_c be the communication channel propagation time, as defined previously.

The RTC Token is sent through the replicator following the last transaction update to ensure that the token is received by the target node after the final update has been received. The time to propagate the RTC Token is therefore the replication latency of the replicator, t_p, plus the communication channel delay, t_c. The return of the RTC Resolution requires another communication channel delay. Thus,

$$L_{cc} = t_p + 2t_c \qquad (4\text{-}2)$$

Synchronous Replication Efficiency

Let us define a comparative measure of synchronous replication efficiency as the ratio of dual write application latency to coordinated commit application latency:

$$e = \frac{L_{dw}}{L_{cc}}$$

where

e is comparative synchronous replication efficiency,
L_{dw} is dual write application latency,
L_{cc} is coordinated commit application latency.

Thus, for $e > 1$, coordinated commits outperform dual writes. For $e < 1$, dual writes perform better.

From Equations (4-1) and (4-2),

$$e = \frac{4n_u t_c + 4t_c}{t_p + 2t_c} = \frac{2(n_u + 1)t_c}{\frac{t_p}{2} + t_c} \qquad (4\text{-}3)$$

Equation (4-3) can be rewritten as

$$e = \frac{2(n_u + 1)}{1 + t_p/2t_c} = \frac{2(n_u + 1)}{1 + 1/p} \qquad (4\text{-}4)$$

where

> p is the round trip communication channel time expressed as a proportion of replication delay time:

$$p = 2t_c/t_p \qquad (4\text{-}5)$$

Figure 4-4 shows replication efficiency e plotted as a function of communication channel time p for various values of transaction sizes n_u. The regions of excellence for dual writes and for coordinated commits are shown.

The values of p and n_u for $e=1$ define the excellence boundary between dual writes and coordinated commits. From Equation (4-4), this relation is

$$2(n_u + 1) = 1 + 1/p$$

or

$$p = 1/(2n_u + 1) \qquad \text{for } e = 1 \qquad (4\text{-}6)$$

This relationship is shown in Figure 4-5.

Figures 4-4 and 4-5 are for the case of parallel read/writes. Remember that the parameter, p, shown on the horizontal axis of Figure 4-4 and on the vertical axis of Figure 4-5, is the ratio of the round trip channel propagation time to the replication latency time. As such, it is a measure of the channel propagation time in terms of latency time. As a ratio, p is dimensionless. For instance, if the round trip channel propagation time is 20 msec., and if the latency time is 50 msec., then p is 0.4. If we reduce the round trip channel propagation time to 10 msec., then p is reduced to 0.2. That is to say, round trip channel time is 20% of the replication latency time.

The efficiency expressions for serial read/writes do not lend themselves to such simple charting. However, the efficiency for serial read/writes will be even better for coordinated commits since dual writes will have the additional application latency of having to wait for the remote reads and writes to complete.

Likewise, these figures are for the case of no simple database operations (operations that may require only a single communication channel round trip such as read/locks, inserts, replacements). To the extent that some updates are simple operations such as these, dual write performance will be better than shown since network traffic is reduced. Our efficiency curves may be modified easily to reflect this situation.

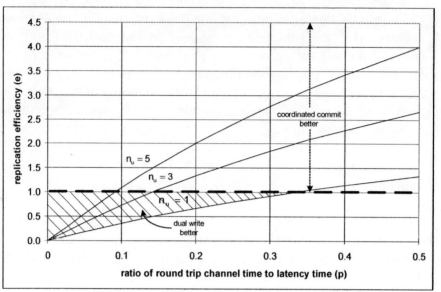

Synchronous Replication Efficiency
Figure 4-4

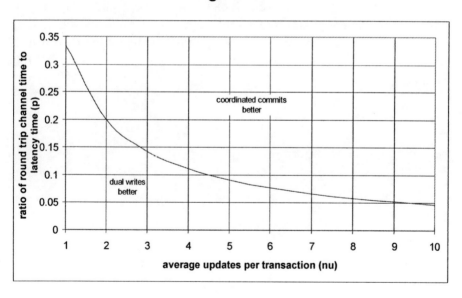

Equal Efficiency (e=1)
Figure 4-5

Scalability and Other Issues

There are other issues to consider when comparing synchronous replication algorithms, including scalability (as it relates to performance) as well as various other algorithm optimizations. We now discuss these issues, including the effects of multiple database copies, communication channel efficiency, and transaction profile.

Scalability

Multiple Database Copies

If there are d database copies in the application network, then an application using dual writes must make d modifications for every database modification required by the transaction. However, under coordinated commits, the application must make only one database modification for each required by the transaction; the other $(d - 1)$ modifications are made by the replication engine and do not affect the application.

Therefore, if other processing is ignored, application latency (which adds to transaction response time) under dual writes will increase with the number of database copies, d, whereas the transaction response time for coordinated commits is relatively unaffected by the number of database copies. Thus, coordinated commits are more scalable than are dual writes.

Communication Channel Efficiency

With dual (or plural) writes, each modification must be sent to all database copies as the modifications are made at the source node. Therefore, this method will send many small messages over the network. The coordinated commit method, on the other hand, has the opportunity to batch multiple change events into a single communication block and thus send several change events within a single block.

For instance, if the application network is making 10 database modifications per second, then 20 messages per second must be sent over the network to each database copy by dual writes (assuming that all modifications require a read followed by a write). Coordinated commits may need to send only one or a few blocks, depending upon the replication latency desired.

If the system is making 1,000 modifications per second, then 2,000 messages per second must be sent to each database copy by dual writes. However, the number of messages sent by coordinated commits, even to achieve small replication latencies, can be one or two orders of magnitude less than this.

As system activity grows, network traffic due to dual writes will grow proportionately. However, network traffic generated by coordinated commits will grow much more slowly as more and more messages are accumulated into a single communication block before they must be sent. Therefore, coordinated commits are much more scalable with respect to communication network loading than are dual writes.

Transaction Profile

Another difference between dual writes and coordinated commits has to do with the transaction profile. In dual writes, the individual database accesses and updates required by a transaction are each delayed as they are sent over the communication channel to the target systems. At commit time, the application must wait for a distributed two-phase commit to complete.

In contrast, with coordinated commits, the source application transaction runs at full speed until commit time, and then pauses while the RTC is exchanged. At the end of this exchange, the application waits for a local (single node) commit to complete. This single node commit should be significantly faster than the distributed two phase commit required to ensure durability on the target systems.

Depending upon the application design, this can be quite advantageous, since the entire application latency is lumped into the commit call time. Furthermore, there may be no slowdown in those applications that support non-blocking (nowaited) commit calls since all of the coordinated commit application time will occur while the application is processing other work.

Read Locks

Coordinated commits and dual writes process read lock operations differently. At times, an application will read a record or row with lock, but not update it (for example, if an intelligent locking protocol is in use). With many implementations of dual writes, the target system record or row will be locked when the source record or row is locked. With coordinated commits, only the source data items are affected – read locks are not necessarily propagated. They only need to be propagated if the data item is subsequently updated. Therefore, not only will dual writes impose more overhead on the network and on the target system, but other transactions may be held up if they are trying to access the target data items that are locked. Thus, coordinated commits should support a higher level of concurrency than dual writes.

Other Algorithmic Optimizations

Of course, some optimizations to the dual write algorithm could be implemented. For example the individual I/O operations could be batched or aggregated and sent together. These optimizations effectively morph the dual write algorithm into a variant of the coordinated commit algorithm with some associated performance and network efficiency gains.

Dr. Bill Highleyman, Paul J. Holenstein, and Dr. Bruce Holenstein

Examples

Geographically Distributed Systems

As an example, consider two systems, one in New York and one in Los Angeles, with the following parameters:

t_p 50 msec.   replication engine processing time, or the time to propagate an update through the replication engine from the source database to the target database, excluding communication channel propagation time.

t_c 25 msec.   communication channel propagation time, or the amount of time required for a message to propagate from the source node to the target node or vice versa over the provided communication channel, including line driver and transmission time.

These parameters result in a value for p (from Equation (4-5)) of 1.0. The resulting efficiency for a given transaction size is, from Equation (4-4),

$$e = \frac{2(n_u + 1)}{1 + 1/1.0} = (n_u + 1)$$

Efficiency as a function of the number of updates for this example is then:

n_u	e
1	2.0
2	3.0
3	4.0
4	5.0

Breaking the Availability Barrier

As is seen, coordinated commits are more efficient for all transaction sizes in the above example. This is a direct result of the communication channel delays. The application latency due to synchronous replication via coordinated commits is, from Equation (4-2), 100 msec. The application latency under dual writes is this value multiplied by the efficiency factor, *e*, in the table above.

Generally, coordinated commits may be more efficient except when the communication times are very small. Note, however, that even at the speed of light, it takes a signal about 25 milliseconds to travel round trip between New York and Los Angeles. Data signals over land lines will take at least twice as long, or about 50 msec. for a New York – Los Angeles round trip. A London – Sydney round trip may take about 250 msec.

Consequently, a generalized statement is that dual writes are appropriate for campus environments with small transaction sizes. On the other hand, coordinated commits are appropriate for wide-area network environments or for large transactions. In addition, an application can be retrofitted to support coordinated commits without any recoding by installing an appropriate data replication facility. This facility usually brings with it auto-recovery of files that have become unsynchronized due to network or node failures, a capability which will probably have to be specially developed for dual writes.

In general, as can be seen from Figures (4-4) and (4-5) and from Equation (4-4), the larger the average transaction or the longer the communication channel propagation time, the more efficient coordinated commits become.

Adding to our rules from Chapters 1 through 3, we have

> **Rule 17:** *For synchronous replication, coordinated commits using data replication become more efficient relative to dual writes as transactions become larger or as communication channel propagation time increases.*

Collocated Systems

If the systems are collocated and interconnected via very high speed channels, the communication channel delay approaches zero; and dual writes may perform better for normal transactions.

However, for long transactions, there will come a point at which coordinated commits will be more efficient. This will be the point at which the processing time for the dual reads and writes at the target node, coupled with whatever communication channel delay exists, exceeds the time to exchange the RTC Resolution. Examples of long transactions are batch streams, box-car'd transactions, and database reorganizations.

To get a feel for this, consider the example of the previous section, but where the communication channel delay, t_c, is 2 msec. instead of 25 msec. In this case, p is .08; and efficiency from Equation (4-4) is

$$e = \frac{2(n_u + 1)}{1+12.5} = \frac{(n_u + 1)}{6.75}$$

Solving this for $e = 1$, we find that replication for transactions of six updates or more will be faster for coordinated commits than it will be for dual writes. For a six-update transaction, the coordinated commit application latency is, from Equation (4-2), 54 msec.

Efficiency Model Extensions

Dual Write Single Round Trip Operations

As pointed out earlier, not all database operations require two communication channel round trips as has been assumed so far. For instance, read/locks, replaces, and inserts require only one round trip – send the operation and receive the completion status. We can extend the above analysis very simply to account for these operations. Instead of there being $4n_u$ communication channel round trips for a transaction, there are $4n_{u2} + 2n_{u1}$ round trips, where

n_{u1} is the number of transaction operations that require one round trip,

n_{u2} is the number of transaction operations that require two round trips,

and $\quad n_u = n_{u1} + n_{u2}.$

Let us define

$$n_u' = n_{u2} + n_{u1}/2 \qquad (4\text{-}7)$$

The previous equations now hold except that n_u is replaced with n_u'. The charts of Figures 4-4 and 4-5 still hold except for the substitution of n_u' for n_u.

Dual Write Serial Updates

If database changes are made serially when using dual writes, then the application must wait for all of the changes to be made not only to its local database but also to the remote database. Let us define t_o as the average database operation time for the particular mix of operations generated by a transaction. Then, in addition to the communication delays, the application will have to wait an additional time of $n_u t_o$ for the remote database operations to complete. The dual write application latency time, originally given by Equation (4-1), now becomes

$$L_{dw} = (4n_u' + 4)t_c + n_u t_o \qquad (4\text{-}1a)$$

and the efficiency relationship given by Equation (4-3) becomes

Dr. Bill Highleyman, Paul J. Holenstein, and Dr. Bruce Holenstein

$$e = \frac{(4n_u'+4)t_c + n_u t_o}{t_p + 2t_c} \qquad (4\text{-}3a)$$

This can be written as

$$e = \frac{2(n_u'+1) + \dfrac{n_u t_o}{2 t_c}}{1 + 1/p} \qquad (4\text{-}4a)$$

where p was defined earlier as,

$$p = 2t_c / t_p$$

If remote database operation time, t_o, is very much less than the communication channel time, t_c, then Equation (4-4a) reduces to Equation (4-4). However, as the communication channel delay t_c approaches zero, Equation (4-4a) approaches

$$e \approx \frac{n_u t_o / t_c}{t_p / t_c} = n_u \frac{t_o}{t_p} \qquad (4\text{-}8)$$

So far as estimating a reasonable value for the average database operation time, t_o, we note the following. Today's high-speed disks (15,000 rpm) have an average random access time of about 7 milliseconds (2 msec. rotational latency, 5 msec. seek time). If the record or row is in disk cache, today's systems require about 0.1 msec. to find it. If buffered writes are used (the write data is checkpointed to another processor and is physically written to disk some time later in the background), then a disk write will take about the same time as a disk read from cache (0.1 msec.), assuming that read and write cache hits are about the same (actually, disk writes might take a little longer due to block splits and index maintenance, but this is ignored here). A well-tuned OLTP application should be achieving something in the order of 80% cache hits. If this is the case, then 80% of read/write accesses will require about 0.1 msec. for cache

Breaking the Availability Barrier

access; and 20% will require about 7 msec. for a physical disk access. This results in an average disk access time of about 1.5 msec.

If we assume typical values for t_o of 1.5 msec. and for t_p of 50 msec., we can see from Equation (4-4a) that serial operations make little difference in efficiency for large communication channel delays. As communication channel delays disappear, the efficiency factor e, rather than going to zero, reduces instead to the value given by Equation (4-8). But for our typical values (and assuming four operations per transaction, or $n_u = 4$), this asymptotic value is only .12. Thus, we can conclude that, in normal cases, dual write serial updates do not significantly affect the performance comparison between dual writes and coordinated commits.

Plural Writes

The above analysis for serial dual write updates assumed that there were only two copies of the database in the application network. But if there are more copies, then the application must wait for each operation to be completed on each remote database in turn. This means that the plural serial write application latency time becomes

$$L_{dw} = (4n_u' + 4)t_c + (d-1)n_u t_o \qquad (4\text{-}1b)$$

where

d is the number of databases in the application network.

The relative efficiency is then

$$e = \frac{2(n_u' + 1) + (d-1)\frac{n_u}{2}\frac{t_o}{t_c}}{1 + 1/p} \qquad (4\text{-}4b)$$

which approaches

$$e \approx (d-1)n_u \frac{t_o}{t_p} \tag{4-9}$$

as communication channel time t_c approaches zero.

If updates are done in parallel, then the original analysis still holds.

Deadlocks

A potential problem with synchronous replication whether it is done via dual writes or coordinated commits, is the possibility of a deadlock. A deadlock occurs when two different applications must wait on locks held by the other. In non-distributed applications, this can occur if the applications are trying to lock two different data items but in different order. Application 1 locks data item A and tries to get the lock on data item B. In the meantime, Application 2 locks data item B and then attempts to lock data item A. Neither can proceed.

This is the standard type of deadlock and can be avoided by an intelligent locking protocol (ILP). Under an ILP, all locks are acquired in the same order. In effect, an application must lock a row or record which acts as a *mutex* (an object which guarantees mutual exclusion). For instance, the application must obtain the lock on an order header before it can lock any of the detail rows or records for that order. In this way, if an application finds a data item locked, all it must do is wait until the owning application has released its locks; and it then can continue on. If an ILP is not being used, then deadlocks can be resolved by one or both applications timing out, releasing their locks, and trying again later at a random time.

Things are not so simple in a distributed system. Although all applications may be following a common ILP, consider an Application 1, running on node X, which locks data item A on its node X. Before that lock is propagated to node Y, Application 2 on node Y locks that same date item A on its node Y. Each application then attempts to lock the same data item on the remote node but

Breaking the Availability Barrier

cannot. A deadlock has occurred even though both applications were following the same ILP.

This deadlock occurred because though both applications were following local ILPs for the nodes they were on, distributing the database and allowing lock access to all copies of all data items in any order on any node means that a global ILP needs to be used. However, there is a solution. One node of the many nodes in the application network must be designated the master node. Locks must first be acquired on the master node before attempting to acquire locks on other nodes. In this way, there is one and only one mutex for each lockable data set.

Failures and Recovery

Dual (Plural) Writes

If a node or network fails in a dual write environment, then the application will have to switch to single node operation. Upon restoration of the node or network, an on-line database comparison and update utility will have to be used to update the remote database (this sort of utility is not normally provided with systems; rather, the entire database must be copied unless a user-written utility is provided). As with asynchronous replication, if an isolated node continues in service, then data collisions may occur during the outage as well as during the recovery process. These collisions will have to be detected and resolved.

Coordinated Commits

<u>Failover</u>

When using coordinated commits, the recovery of user services following a node or network failure is almost identical to that for asynchronous replication, as described in Chapter 3, "*Asynchronous Replication*." In the event of a node failure, users may be switched from the failed node to the surviving node. If necessary, full operation resumes with the load shedding of non-critical functions. There is one important difference.

With asynchronous replication, transactions in the replication pipeline from the failed node are lost. With synchronous replication, no data is lost. Rather, one or more transactions may be held in an uncertain state. That is, if a node owning an outstanding transaction should fail after its target nodes have acknowledged its prepare-to-commit command or its RTC token but before the target nodes have received the commit command, then the target nodes do not know whether the source commanded either a commit or an abort or whether the source failed before it could do either. In this case, the target transactions are hung but are not lost. What is done with these transactions is application dependent, but at least the data is not lost.

Should the failure be a network failure, then it may be decided to switch users to a node which is not isolated. An alternative decision may be to allow the isolated node to continue independently.

Restoration

In any event, during the outage, transaction updates will queue at the active nodes and will be sent later to the downed or isolated node when it is returned to service. At this time, replication proceeds as asynchronous replication to bring the remote database into synchronization. If isolated nodes continue in service so that transactions are being replicated in both directions, collisions will have to be detected and resolved as described in Chapter 3.

When the Change Queues have been drained to the point that the replication latency is deemed to be acceptable, users can be re-switched to their home nodes; and synchronous replication can be restarted. Synchronous replication messages will simply queue behind the remaining asynchronous messages until those asynchronous messages have been processed, at which time the application returns to purely synchronous replication.

What's Next?

In Chapters 1 and 2, we viewed availability with respect to systems which require repair and which are returned to service the instant the repair is complete. However, in the real world of redundant systems, a system seldom needs repair to return it to service. Rather, faults are transient; and the system needs to be recovered rather than repaired. Chapter 5, *"The Facts of Life,"* discusses this twist on system availability.

There are three major categories of failures with that to be concerned relative to asynchronous replication:

- the source node fails.
- the network fails.
- the target node fails.

Of course, in a bi-directional replication environment, a single-node failure represents both a source node failure and a target node failure because the failed node had been performing both functions.

There are two recovery points of interest in a data replication environment. One is the immediate recovery of processing functions following a failure. We will call this *failover*. The other is the return to service of the failed component once it is repaired. We will call this *restoration*.

Chapter 5 - The Facts of Life

In the first four chapters of this book, we focused on some basic concepts and applied these concepts to the development of architectures that offered significantly enhanced availability. The model that we used was that of a system that comprised multiple identical subsystems. Of these subsystems, s were spares; and the system could tolerate the failure of any s subsystems. However, the failure of $s+1$ subsystems might cause the failure of the system. In the event of a system failure caused by $s+1$ subsystem failures, the system was immediately restored to service upon the repair of one of the failed subsystems.

This model is accurate for replicated and split systems. For instance, if a split system comprises two nodes, and should both nodes fail, then service is restored as soon as one of the failed nodes is repaired and brought back into service (assuming that no database recovery is required before the system can be used).

However, within a failed node, things are not so simple. Once the requisite number of subsystems have been repaired and are operational, the node often must be recovered before it can be returned to service. This may require a variety of actions taking several hours.

In this chapter, we take a look at the impact of system recovery which may have to follow system repair. But first, let us review what we have done to date.

Dr. Bill Highleyman, Paul J. Holenstein, and Dr. Bruce Holenstein

A Review of Availability

In our previous chapters, we developed the general availability relation

$$A \approx 1 - f(1-a)^{s+1} \approx 1 - f\left(\frac{mtr}{mtbf}\right)^{s+1} \quad (5\text{-}1)$$

where

- A is the availability of a system comprising similar redundant subsystems.
- a is the availability of a subsystem.
- s is the number of spares provided in the system. That is, any s subsystems may fail; and the system will continue to operate.
- f is the number of failure modes, or the number of ways that $s+1$ subsystems will fail in such a way that a system outage will result.
- mtr is the mean time to repair a subsystem.
- $mtbf$ is the mean time before failure for a subsystem.

Availability is the probability that the system will be operational and is therefore

$$A = \frac{MTBF}{MTBF + MTR} \quad (5\text{-}2)$$

where

- $MTBF$ is the mean time before failure for the system.
- MTR is the mean time to repair the system (or, more accurately, the mean time to *restore* the system to service)

We discussed the characterization of availability in terms of 9s; e.g., an availability of .9999 is called four 9s. We showed that adding a backup doubles the nines, which is the basic power of fault-tolerant systems. We discussed the relative efficiencies of methods to keep replicated databases in exact synchronization. We also showed that splitting a system into several independent nodes using data replication can provide substantial improvement in availability at little or no additional cost.

However, systems of multiple redundant processors suffer from a plurality of failure modes which can reduce system availability by one or two 9s. We showed that by careful allocation of processes to processors, we can substantially reduce this impact.

We so far have applied these concepts to hardware failures as if hardware failures were the predominant cause of system outages. This has been the mechanism that has allowed us to demonstrate the application of these concepts. However, this is not real life. Recent experience indicates that only a very small portion of system outages are caused by dual hardware failures. The rest are caused by complex interactions between hardware failures, software failures, operational errors, and environmental faults.

Adding to the rules which we derived in our earlier chapters, we have

Rule 18: *Redundant hardware systems have an availability of five to six 9s. Software and people reduce this to four 9s or less.*

The work to date is no waste of time. The concepts are valid. But in order to apply these concepts to real-life availability, we have to understand better what is going on.

Why Do Computers Stop?

Jim Gray is unquestionably one of the key contributors to fault-tolerant computing. In 1985, he published "*Why Do Computers Stop and What Can We Do About It*,"[1] a defining paper on the real causes of system outages. This paper formed the basis for his chapter on "*Software Fault Tolerance*" in his subsequent book <u>Transaction Processing: Concepts and Techniques</u>,[2] which sets forth the basic principles underlying fault-tolerant systems.

Gray's footprints can be found all throughout this chapter. Rather than citing each reference, let us suggest that you read his paper, which can be found at <u>http://www.cs.berkeley.edu/~yelick/294-f00/papers/Gray85.txt</u>.

Though Gray's paper was published in 1985, it remains strikingly applicable today. Little has changed except that hardware has become somewhat more reliable, as he predicted. We summarize his observations next with some updating comments.

Gray studied 166 unscheduled system outages reported to Tandem Computers over a seven-month period. The outages covered over 2000 systems. Note that this equates to a system availability of .99993, assuming an MTR of four hours (see later). It confirmed Tandem's claim to availabilities of about four 9s.

About one-third of the reported outages related to "infant mortality" problems, defined as recurring problems which were later fixed. If these were removed, only 107 system outages were reported during this period. The outages were characterized as follows (the Standish results are described later):

---

[1] Gray, J.; "*Why Do Computers Stop and What Can We Do About It?*" 5th <u>Symposium on Reliability in Distributed Software and Database Systems</u>; 1986.
[2] Gray, J.; et al.; <u>Transaction Processing: Concepts and Techniques</u>, Morgan Kaufmann; 1993.

	Gray 1985	Standish 2002
People	42%	38%
Software	25%	28%
Hardware	18%	17%
Environment (power, a/c, etc.)	14%	17%
Unknown	1%	-

Contributors to Tandem NonStop Outages
Table 5-1

There are some caveats that must be made about these observations. First, they probably don't include many outages caused by application software faults or environmental problems. Customers don't generally report these. Second, they probably don't include all operator errors. Operators don't always report their goofs. Finally, they don't include scheduled down time.

Gray points out in his 1985 paper that "....hardware will be even more reliable due to better design, increased levels of integration, and reduced numbers of connectors." In fact, this forecast has turned out to be quite correct. Experience now indicates that outages due to dual hardware failures represent less than 5% of all outages.

Gray goes on to say, "....the trend for software and system administration is not so positive. Systems are getting more complex." A recent study by Standish Group[26] supports this observation. Clearly, not much has changed since 1985. Citing a NonStop server reliability of .9998 (classic Tandem systems are now known as NonStop servers following the acquisition of Tandem by HP), Standish's measurements of the causes of system outages are also shown in Table 5-1. They are amazingly consistent with Gray's findings. Interestingly, the Standish study also indicates that network failures happen more often than server system failures; and outages caused by applications swamp server system failures by almost four to one.

---

[26] Standish Group, <u>VirtualBEACON</u>, Issue 244; September, 2002

Note that about 40% of all outages seem to have been caused by human error. However, this isn't as bad as it may look. Forty-five outages per year over 2000 systems represent one human error per system every 44 years. Don't we all wish that we were that accurate?

An interesting insight into operator errors has been nicely phrased by Wendy Bartlett of HP. It recognizes the stress factor which accompanies an unexpected failure:

Rule 19: (Bartlett's Law) *When things go wrong, people get stupider.*

Because of this human characteristic, it is very important to have a recurrent training program to ensure that operations staff can react properly in an emergency and that the established procedures are correct. Ideally, failure situations should be simulated periodically so that the operations staff can practice proper recovery. The more automatic their reactions become, the less stupid they will appear when the heat is on. This leads to the following corollary to Rule 19:

Rule 20: *Conduct periodic simulated failures to keep the operations staff trained and to ensure that recovery procedures are current.*

Today, the numbers may be a little different. In fact, they may be different each year because the sample size is too small. However, all indications are that the story these numbers tell is the same:

Rule 21: *System outages are predominantly caused by human and software errors.*

Rule 22: *Seldom does a recovery entail hardware repair. It entails a reload of the system.*

Let us consider further these Rules 21 and 22.

Some Definitions

Let us first define some terms:

A *fault* is a lurking incorrectness waiting to strike. It may be a hardware or software design error, a hardware component failure, a software coding error, or even a bit of human ignorance (such as an operator's confusion over the effects of a given command).

A *failure* is the exercise of a fault. Failures in themselves do not cause outages. Fault-tolerant systems are designed to survive any single failure.

A *failover* is a recovery from a failure by switching to a backup component.

A *failover fault* is a failure of a failover.

An *outage* is a denial of some service to part or all of the user community. Outages can range from unacceptable response times to total system unavailability.

A *trigger* is an initial failure that begins an event sequence that leads to a subsequent failure from which the system cannot recover. The result is a system outage.

A *repair* is the return to service of a component which has experienced a failure.

A *recovery* is the return to service of a system which has experienced an outage.

Note that the *repair* of a failed subsystem does not necessarily result in the *recovery* of its system.

Dr. Bill Highleyman, Paul J. Holenstein, and Dr. Bruce Holenstein

Triggered Outages

Our availability analysis so far has assumed (in today's fault-tolerant environment) that an outage is caused by two failures *and* that these two failures are independent (i.e., one failure does not induce the other). Although simplistic, this view of things has led us to several important concepts in availability.

We now see that dual failures which cause outages are hardly independent at all. Rather, one random failure often leads to a directly related second failure, which causes the outage. For example,

- a disk unit fails, and the good disk is erroneously pulled for replacement.

- a critical process aborts, and a bug in its backup checkpointing procedures causes the backup to fail as well.

- An error (a fault) in the operations manual is followed by the system operator and results in an outage.

In general, a failure occurs and must be corrected but otherwise does not seriously affect system operation. But then that failure triggers another failure and a system outage results.

This leads to another interesting observation. You may have noted that new systems seem to be less reliable than established systems. A system seems to "burn in" with time. How can this be?

New systems are subject to continuous change. Functional errors are corrected, bugs are worked out, and enhancements are made. Each change carries with it the potential for further errors which may act as outage triggers. As the system matures, changes become less frequent; and system reliability improves. As Carl Neihaus of HP has said,

Rule 23: (Niehaus' Law) - *Change causes outages.*

The Impact of Failover Faults

In Chapter 1 of this series, we showed that the probability of an outage for a system configured with one spare and with randomly distributed processes is

$$F \approx \frac{n(n-1)}{2}(1-a)^2 \qquad (5\text{-}3)$$

where

      F is the probability of a system outage.
      n is the number of subsystems in the system.
      a is the probability of a subsystem failure.

That is, a subsystem will fail with a probability of $(1-a)$. A pair of subsystems will fail with a probability of $(1-a)^2$, and there are $n(n-1)/2$ ways in which two subsystems might fail.

This relation assumes that the outage is caused by the independent failures of two subsystems, whether those failures are caused by hardware or by software. Implicit in Chapters 1 and 2 was the assumption that critical process pairs were trusted and did not fail. A system outage was caused by a dual hardware failure, which took down a critical process pair or which denied a critical process pair access to needed data. We now relax that assumption and allow a subsystem to fail due either to a hardware failure or to a software failure. Furthermore, we realize that the *failover* mechanism which should recover from that failure by invoking the hardware or software backup may fail itself and thereby create a *failover fault*.

Thus, we now know that not all outages are caused by dual independent failures. With a probability of p, a *single* subsystem failure, whether it be due to hardware or software, will experience a failover fault that leads to a system outage. Only $(1-p)$ of all outages are caused by dual subsystem failures, where

      p is the probability that a failover attempt will fail (a failover fault).

We now have two failure modes for a system:

- two subsystems have failed.
- one subsystem has failed, and the failover has failed.

Furthermore, there are two components to system restoration – subsystem repair and system recovery. Let

r be the mean time to repair a subsystem.
R be the mean time to recover a system.

Also, remember that the average time to return the system to service when there are two failed subsystems is $r/2$ (see Rule 7 in Chapter 1).

Let us consider each of the two failure modes listed above.

(1) <u>Two subsystems have failed.</u>

The probability that two subsystems will fail is given by Equation (5-3). $(1-p)$ of all outages will be caused by a dual subsystem failure. Furthermore, the effective system repair time is the time to repair one of the subsystems, $r/2$, plus the time to recover the system, R, giving a total repair time of $r/2+R$ rather than just $r/2$. Since down time is proportional to repair time, the failure probability for this mode is increased by a factor of $(r/2+R)/r/2$:

outage probability due to dual failures

$$\approx \frac{r/2+R}{r/2}(1-p)\frac{n(n-1)}{2}(1-a)^2 \qquad (5\text{-}4a)$$

(2) <u>One subsystem has failed, and the failover has failed</u>

The probability that a particular subsystem will fail is $(1-a)$. There are n ways in which a system can experience a single subsystem failure. Thus, the probability that one subsystem in

Breaking the Availability Barrier

the system will fail is $n(1-a)$. Of all outages, p are caused by single subsystem failures followed by a failover fault.

Should an outage be caused by a failover fault, there is no need for a subsystem repair since no more than one subsystem is down. Thus, subsystem repair time, r, has been replaced by system recovery time, R. Therefore, the probability of failure for this mode is modified by a factor of R/r:

outage probability due to failover fault

$$\approx \frac{R}{r} pn(1-a) \qquad (5\text{-}4b)$$

Recognizing that the system failure probability F is the sum of the above two probabilities, and assuming that p is very much less than 1, then

$$F \approx \frac{r/2 + R}{r/2} \frac{n(n-1)}{2}(1-a)^2 + \frac{R}{r} pn(1-a) \qquad (5\text{-}4c)$$

These relationships are good approximations describing the impact of failover faults on system availability. They are proven formally in Appendix 3, "*Failover Faults.*" The following observations are made about these relationships:

A Better Value for Subsystem Availability

So far, we have assumed that system availability A is about four 9s, given a subsystem availability a of .995. Now we can deduce a better value for a from Equation (5-4c).

Let us consider an 8-processor system (n = 8) with an availability A of four 9s (F = .0001). We assume a repair time r of 24 hours, a recovery time R of 4 hours, and a failover fault probability of 1%. Solving Equation (5-4c) for a, we find that a better value for subsystem availability a is .9986, more than three times better than we previously assumed. This is because we now recognize that some outages are caused by failover faults and not by dual subsystem failures.

Effect of Failover Faults on System Availability

We can calculate from Equations (5-4) that a 1% failover fault rate causes 20% of all system down time under the previous example. Note from Equation (5-4b) that this is directly affected by system recovery time R. To the extent that we can reduce recovery time, we can minimize the effects of failover faults.

Effect of Failover Faults on Effective Subsystem Availability

A further insight into the impact of failover faults on system availability is gained as follows.

We can rewrite Equation (5-4c) as

$$F \approx \frac{r/2 + R}{r/2} \frac{n(n-1)}{2} (1-a)(1-a') \qquad (5\text{-}5)$$

where

$$a' = a - \frac{R}{r/2 + R} \frac{p}{n-1} \qquad (5\text{-}6)$$

Note that a' is less than a, being reduced by a subtractive term. Thus, comparing Equation (5-5) to Equation (5-4a) (and ignoring the $1-p$ term as being very close to one), we can make an interesting interpretation. The failure of the first subsystem will occur with a probability of $(1-a)$ as expected. *However, once one subsystem has failed, the system then behaves as if it comprises n-1 remaining subsystems with decreased availability a'.*

A simple example will serve to illustrate this. Let us consider a 4-processor node ($n = 4$) comprising subsystems with an availability, a, of .9986 (as calculated above). Subsystem repair time, r, is 24 hours and system recovery time, R, is four hours. If the probability of a failover fault, p, is 1%, then the effective subsystem availability, a', following a single subsystem failure, is reduced from .9986 to .9978. Failure probability has increased from .0014 to .0022. Under the above parameter assumptions, a 1% chance of a failover fault makes the system 60% less reliable following a single subsystem failure!

This leads to the following rule:

Rule 24: *Following the failure of one subsystem, failover faults cause the system to behave as if it comprises n-1 remaining subsystems with decreased availability.*

Note also that as recovery time R decreases, Equation (5-6) shows that the reduced subsystem availability a' improves and approaches the subsystem availability a. Thus, reducing recovery time reduces directly the impact of failover faults on system availability.

Effect of Failover Faults on System Splitting

In Chapter 2, we showed that splitting a system into k nodes improved system reliability by at least a factor of k:

$$\text{Split System Reliability Improvement} = k\frac{n-1}{n-k} > k \quad (5\text{-}7)$$

where

k is the number of nodes into which the system is split.
n is the number of processors in the system.

Thus, Equation (5-7) predicts that the reliability improvement achieved by system splitting is always greater than k. However, such a gain is compromised when we have the possibility of failover faults. As shown in Chapter 11, "*Failover Faults*," this relation then can be shown to become

$$\text{Split System Reliability Improvement} = k\frac{(n-1)+x}{(n-k)+kx} \quad (5\text{-}8)$$

where we can think of x as an exasperation factor. If x is very small, then Equation (5-8) approaches Equation (5-7); and we have the split system advantage we are seeking. However, if x is very large, then Equation (5-8) approaches one; and the reliability advantage of system splitting disappears.

x is given by

$$x = \frac{R}{r/2 + R} \frac{2p}{1-a} \quad (5-9)$$

Note that as recovery time R becomes smaller, x becomes less significant; and we recover our availability advantage provided by system splitting. Thus,

Rule 25: *The possibility of failover faults erodes the availability advantages of system splitting (see Rule 9).*

As an example, from Equation (5-7) we expect that splitting a 16-processor system into four 4-processor nodes will give us a reliability advantage of a factor of 5. But suppose that we have a failover fault probability, p, of 1%. Furthermore, assume that subsystem availability, a, is .9986, that subsystem repair time, r, is 24 hours, and that system recovery time, R, is four hours. In this case, from Equations (5-8) and (5-9), the availability advantage of system splitting decreases from a factor of 5 to a factor of 2.8.

The Golden Rule – Reduce Recovery Time

We have seen the importance of minimizing recovery time to improve system availability. A further insight can be gained by noting that about 20% of CPU halts are caused by hardware failures, and about 80% are caused by software faults or human errors. Thus, only about 4% of system outages (20% x 20%) are caused by dual hardware failures.

The remaining 96% of outages are caused by no more than one hardware failure combined with a software fault or a human error.

Breaking the Availability Barrier

These outages do not require a repair to return them to service. They only require a recovery:

Rule 22 (restated): *A system outage usually does not require a repair of any kind. Rather, it entails a recovery of the system.*

We now have seen that recovery time is a predominant factor in system availability:

- The time to recover from most outages is recovery time rather than repair time. Therefore, any reduction in recovery time is *directly* reflected in system availability.

- Minimizing recovery time helps negate the effect of failover faults on system failure rate.

- Minimizing recovery time helps maximize the availability advantages of system splitting.

This leads to what might be considered the golden rule of availability:

Rule 26: (The Golden Rule) - *Design your systems for fast recovery to maximize availability, to reduce the effect of failover faults, and to take full advantage of system splitting.*

What can we do to reduce recovery time?[27] The recovery process is quite complex. It may entail

- realizing that a problem has occurred.
- diagnosing the cause of the problem.
- deciding what to do.
- obtaining permission to follow a course of recovery action.
- collecting diagnostic data (such as a processor memory dump).

---

[27] For a description of on-going research into ways to minimize recovery time via Recovery-Oriented Computing (ROC), see A. Fox, D. Patterson, "*Self-Repairing Computers,*" Scientific American; June, 2003.

- cold-loading the system.
- restarting software subsystems.
- restarting the applications.
- restarting the network.
- perhaps recovering the database.

A typical recovery procedure takes anywhere from an hour to several hours. Four hours is a pretty good guess at an average recovery time.

As can be seen from the above list, recovery time isn't minimized by simply building our applications for quick recovery, although this, of course, is very important. Efficient recovery also requires

- good operator training.
- efficient decision making by the entire management team.
- well-documented recovery procedures for
    o the system.
    o the application.
    o the database.
    o the network.

Rule 27: *Rapid recovery of a system outage is not simply a matter of command line entries. It is an entire business process.*

The Importance of Restore Time

Let us define *restore time* as the time required to return a system to service. Restore time may entail a hardware repair and probably requires a system recovery. The *MTR* term in Equation (5-2) should really stand for *mean time to restore*.

The importance of restore time (which can also be called down time) so far has only been considered as a factor in availability. To the extent that one can reduce restore time, one can increase system availability.

Breaking the Availability Barrier

However, restore time has a far greater importance in its own right. It may be that the average cost per down time hour is a user's measure of the value of reliability. But this cost is also a function of the length of a down time interval. As down time grows longer, the outage cost may increase. Customer annoyance may give way to customer anger, then lost sales, then lost customers.

Down time is even more profound for safety-critical functions. For instance, 911 operators will probably feel that a five-second outage is an annoyance. But a five-minute outage can mean a death due to cardiac arrest or a building being burned to the ground.

Fault-tolerant systems are full of outages that are seen not as outages at all but as very brief periods of perhaps degraded response times. These are the normal outages caused by subsystem failures that are recovered in seconds. For instance, in many fault-tolerant systems, recovery is accomplished by failing over to a backup process or by resubmitting a transaction to a surviving server process. It is the severe outages requiring a system recovery that currently limit fault-tolerant systems to four 9s.

Since down time is predominantly recovery time for fault-tolerant systems, we perhaps can add a nine to the current fault-tolerant availability if we pay more attention to designing systems and business processes for fast recovery. An excellent example of this is a major corporation which runs a variety of applications on several HP NonStop servers. Every one of their applications is backed up by another system that either normally runs other applications or is a dedicated backup. Recovery time for many of their replicated applications is about two minutes. A typical NonStop system has an availability of four 9s and an average recovery time of four hours. By cutting the recovery time from four hours to two minutes – over a 100:1 improvement in restore time – this corporation has added two 9s to its availability and thus has created an availability in excess of six 9s.

Dr. Bill Highleyman, Paul J. Holenstein, and Dr. Bruce Holenstein

What's Next?

We have developed in the last four chapters several concepts regarding availability and the related parameters of mean time before failure and mean time to repair. We have shown that splitting a system into several independent nodes can significantly improve the reliability of a system, and have discussed how these nodes can remain synchronized. We have discussed the impact of system recovery on these parameters.

In Chapter 6, "*RPO and RTO*," we look at two important objectives for high availability systems and discuss how various replication architectures map into these objectives.

Chapter 6 - RPO and RTO

In the previous chapters, we have discussed how system availability can be significantly enhanced by replicating a system or by breaking it up into several smaller independent cooperating nodes. We have discussed to some extent the various failure modes of distributed systems, the immediate recovery from these failures, and the ultimate restoration of the full system to service following the repair and recovery of a failed node.

This chapter explores two important questions relative to failures in a distributed system that must be considered when deciding how to replicate or split a system:

- How much data can be lost due to a node or network failure? This is called the Recovery Point Objective, or RPO.

- How much time can be lost due to a node or a network failure? This is called the Recovery Time Objective, or RTO.

Replicating the System

Fundamental to the achievement of high availability in distributed systems is the provision of one or more similar nodes, either locally or at remote sites, that can provide the application services to all users in the event that a node in the distributed system becomes unavailable for any reason. If nodes are geographically dispersed, then the system is protected also from natural or man-made disasters.

Of course, replicating a system involves much more than just the provision of computing capacity at other sites. It requires adequate staffing at all sites, the network facilities to switch remote users from one site to another, the transfer of the current state of the application databases, and solid procedures for the recovery process.

In this chapter, we focus on methods for making up-to-date copies of the application database available at multiple sites. We will explore different methods for doing so and will discuss the recovery process for bringing up a consistent database at a surviving node following the failure of another node so that application services can be restored to all users who had been serviced by the failed node. Specifically, we will evaluate the time it takes to recover from a node failure and the risk of data loss caused by such a failure.

RTO and RPO

When considering the true and total costs of a node failure, two important factors are the time that the users of the system are denied service and the amount of data, if any, that may have been lost due to the failure. Of course, associated with a recovery are many other costs, such as the movement of personnel from the failed site to the backup site. However, we will focus here on the issues of denied service and lost data.

When deciding how nodes in a distributed system will back each other up, the amount of tolerance that a business has to recovery time and lost data should be established as a pair of goals, or objectives, that the system must achieve in the event of a primary system loss. This tolerance can be established as Recovery Point and Recovery Time Objectives[28]:

- The Recovery Point Objective, or *RPO*, is a measure of how much data loss due to a node failure is acceptable to the business. A large RPO means that the business can tolerate a great deal of lost data.

---

[28] LaPedis, R.; "*Will Enterprise Storage Replace NonStop RDF*," The Connection, Volume 23, Issue 6; November/December, 2002.

- The Recovery Time Objective, or *RTO*, is a measure of the users' tolerance to down time. A large RTO means that users can tolerate extensive down time.

These two objectives are not closely related – they may both be almost zero, they both may be large, or one may be small but the other large.[29] Various examples of the needs of different applications are shown in Figure 6-1. For instance, a stock exchange trading system must be brought back to life very quickly and can lose no data. Since the price of the next trade depends upon the previous trade, the loss of a trade will make all subsequent transactions wrong. In this case, the RTO may be measured as a few minutes or less, but the RPO must be zero.

On the other hand, a critical monitoring system such as those used by power grids, nuclear facilities, or hospitals for monitoring patients must have a very small RTO, but the RPO may be large. In these systems, monitoring must be as continuous as possible; but the data collected becomes stale very quickly. Thus, if data is lost during an outage (large RPO), this perhaps impacts historical trends; but no critical functions are lost. However, an outage must end as quickly as possible so that critical monitoring can continue. Therefore, a very small RTO is required.

A Web-based store must have an RPO close to zero (the company does not wish to lose any sales or, even worse, acknowledge a sale to a customer and then not deliver the product). However, if shipping and billing are delayed by even a day, there is often no serious consequence, thus relaxing the RTO for this part of the application.

A bank's ATM system is even less critical. If an ATM is down, the customer, although aggravated, will find another one. If an ATM transaction is lost, a customer's account may be inaccurate until the next day, when the ATM logs are used to verify and adjust customer accounts. Thus, neither RPO nor RTO need to be small.

---

[29] LaPedis, R.; "*RTO and RPO Not Tightly Coupled*," Disaster Recovery Journal; Summer, 2002.

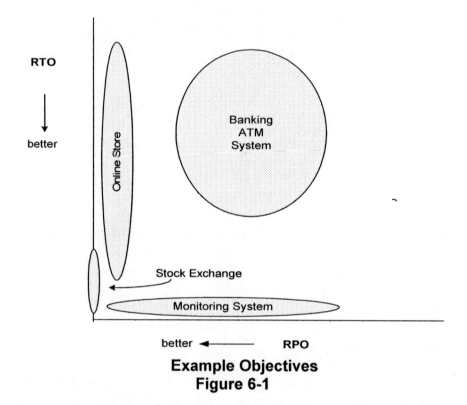

Example Objectives
Figure 6-1

Once a company decides what RPO and RTO are applicable to an application, the method for backup and recovery of that application becomes much more evident. The remainder of this chapter relates various methods for backup and recovery to their RPO and RTO characteristics.

Replicating the Application Data

Immediately recovering services to users on a failed node requires that a current copy of the application database be available on at least one other node accessible to all nodes in the system, and that users be switched to a surviving node. Providing a replicate of the application

Breaking the Availability Barrier

data at another node can be done in many ways. Classically, magnetic tape has been used to periodically take a snapshot of the database so that it can be loaded onto the backup system. This is still the most common means for database recovery as of the writing of this book. Although this technique results in a large RPO (many hours of data may be lost) and a large RTO (it may take hours to days to restore service), tape backup is relatively inexpensive.

However, several data replication techniques allow a remote database to be kept in near synchronism with the primary database, thus reducing the RPO to a very small value (a few seconds or less of lost data). In some cases, exact synchronism can be achieved, thus reducing RPO to zero. These techniques also lend themselves to very fast recovery, or small RTO, measured in minutes or even in seconds.

Various data replication techniques are shown in Figure 6-2 and are described next. To simplify the discussion, we will first consider replication between two systems. In some cases, one might be a primary system and the other a passive backup system. In other cases, both systems might be active with each serving as a backup system to the other.

These techniques are extendable to application networks comprising several independent but cooperating nodes. The implications of multiple-node application networks follow the description of the basic techniques.

No Replication

The use of tape to back up a system requires no data replication facility. Such backups tend to have long recovery times and may have a probability of significant data loss. The backup system is not contributing to application processing.

Dr. Bill Highleyman, Paul J. Holenstein, and Dr. Bruce Holenstein

No Replication, Periodic Backup Only

The classic method for restoring a database at a backup site is via magnetic tape. Periodically, a snapshot of the database, or of the changes made to the database since the last snapshot, are written to tape. If the snapshot is to represent a consistent view of the database, the application must be paused during the snapshot. If it is not paused, the snapshot represents an inconsistent view of the database since it will show updates partially made by some transactions which may or may not be committed.

In the event of a failure, if the latest backup tapes have not been applied to the backup system, they must be recovered from the tape storage site and moved to the backup site. They must then be loaded before the backup site can take over operations. There is some probability that a tape will be unreadable. This will require the retrieval of an alternate backup tape if one exists. If multiple backup tapes have not been created, then a significant amount of data will be lost if a tape is unreadable.

If the backup tapes have already been applied to the backup system, then the database reload is complete as far as it can go. All database updates between the last backup and the time of the failure are lost. Thus, not only does this backup method have a potentially long recovery time (large RTO), but it is also susceptible to a large amount of data loss (large RPO).

No Replication, Periodic Backup with Audit Trail

In some systems, a continuing audit trail of database changes made since the last snapshot is written to tape. Playing back the audit trail against an inconsistent snapshot can restore a recovered database to a consistent state. Under proper procedures, the snapshot copies are applied to the backup system as soon as practicable. Then, should an outage occur, only the audit trail tapes since the last snapshot (at least, those that can be made available) need be applied to provide a reasonably up-to-date database for the backup system.

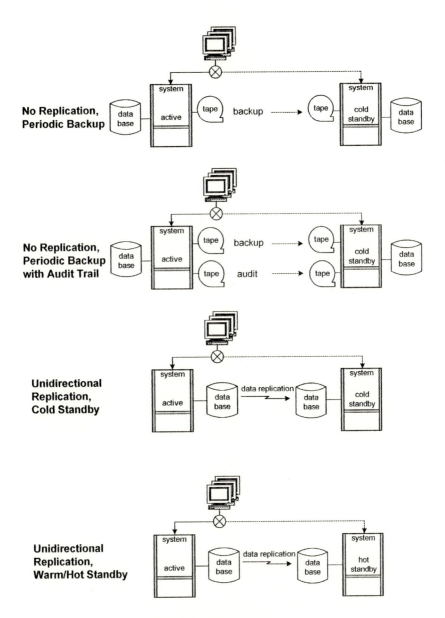

**Replication Methods
Figure 6-2**

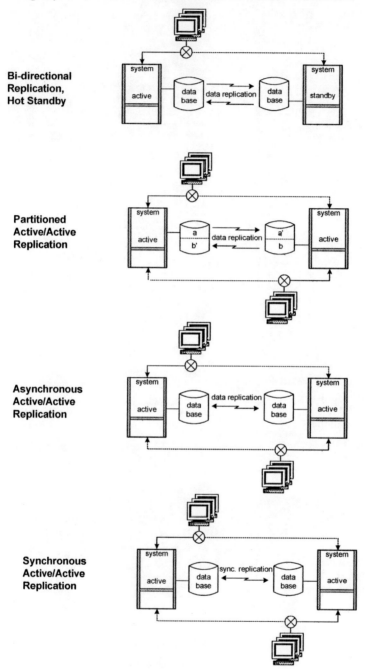

Replication Methods
Figure 6-2 (cont)

Playing back the audit trail tapes may extend the recovery time (RTO) if snapshot tapes have to be loaded first, but doing so can dramatically reduce the amount of lost data (RPO).

Unidirectional Replication

With unidirectional replication, the application is only running on the active system. The backup system may use the replicated data for read-only purposes but cannot be actively updating that data.

Unidirectional Replication – Cold Standby

The simplest form of data replication feeds a backup database in near real time with updates that are being made to the active database. Except for data replication activity, the backup system has no participation in the application. There are no application processes running on the backup system, though the system may be used for other work. Thus, it is called a cold standby.

Unidirectional Replication – Warm or Hot Standby

The same data replication procedure can be used with a warm or hot standby. Both warm and hot standbys have all associated applications up and running. However, a warm standby has opened files only for read access. If it is to take over, it must reopen these files for full access. A hot standby has all files open for full access and is available to immediately take over in case of a primary failure. In such cases, updates are still replicated from the active system to the backup database. However, until the backup applications are requested to take over, they are not making any changes to the database. If they are doing any work, they are only executing query and reporting functions against the database.

Bi-Directional Replication

A primary system with a hot backup system may also be configured with bi-directional replication. In one configuration, it will behave exactly as if it were a unidirectionally replicated hot standby system as described above, except that now it is rather straightforward to switch active/standby roles. This can be particularly advantageous for testing system recovery and keeping operations personnel current in switchover procedures via periodic training.

To switch systems for testing purposes, all that is necessary is to switch users to the standby system. Since all applications are up and running, the only impact on users will be the short switchover time. There will be no data loss as there might be if the primary system failed (the time required for the replication engine to drain its primary system workload will usually be sub-second). However, as the users now continue with their normal activities, the old primary system acting as the new standby system will continue to have its database kept in synchronism with the new active system's database. A return to the original configuration can be accomplished at any time by simply switching users.

Active/Active Replication

A major advantage of bi-directional replication is that the full capacity of both systems in the network is available for application processing. Thus, the reliability and disaster tolerance afforded by replicating systems can be achieved with a much smaller complement of equipment. With no replication or with unidirectional replication, only half of the system capacity is available for application processing (though it may be available for query and reporting or for running other applications that do not affect the replicated application's database).

If applications are running in the backup system, there is no reason why they can't be actively engaged in providing the same services as the primary system. Each system can be supporting its own community of users, be making its own updates to its local database, and be replicating these updates to the remote database. Such distributed applications are called *active/active* applications.

However, there are some serious problems with active/active replication. One is ping-ponging, or the replication by a system of an update back to that system from which it was received. Provisions must be made to avoid ping-ponging.[30]

Another problem is data collisions. A data collision occurs when two users update the same data item on different copies of a database. Since the same data item now has different values in different instances of the database, typically only data content or application-knowledgeable reconciliation procedures are capable of returning the database copies to a consistent state. That is, the correct state of the affected data item depends upon how it is used by the application. In some cases, the data collision may be safely ignored, or it may be resolved automatically by the replication engine. In others, the collision needs to be resolved manually.

Partitioned Active/Active Replication

Data collisions can be avoided by logically *partitioning* the database and the processing activity so that each system will still have a full copy of the database, but the users at each system will update only the database partition that their system owns. These updates can then be replicated to the other system with no fear of data collisions.

For instance, it may be that only the sales office that services a customer can make updates to that customer's records. A cell phone service provider might send call records to a specific system based on the first digit of the calling or called number.

Partitioning might even be done on a time basis. The ownership of the database might be rotated to give different systems an opportunity to process requests which they have been queuing, or a brokerage's database might be transferred during the 24-hour day to offices which are currently open around the world.

---

[30] Strickler, G.; et al.; "*Bi-directional Database Replication Scheme for Controlling Ping-Ponging,*" United States Patent 6,122,630; Sept. 19, 2000.

Asynchronous Active/Active Replication

If data collisions are not deemed to be a serious problem, then both systems can be actively processing all transactions; and no partitioning is required. This may be the case, for instance, if users are geographically segregated and if the risk of data collisions is small. Alternatively, if the cost of resolving data collisions is small, or if data collisions can be resolved automatically, asynchronous active/active replication may also be appropriate.

Synchronous Active/Active Replication

In many applications, partitioning will not work or may not be feasible. Any transaction may possibly affect any data item in a database, and resolving data collisions may simply be too difficult or require too much manual intervention to use asynchronous active/active replication. Both systems must cooperate to ensure that a transaction they own will not cause a data collision with a transaction owned by the other system.

This can be accomplished by ensuring that a transaction can lock throughout the network all of the instances of a data item which are affected by that transaction. It can then update all data item instances before releasing the locks. We call this *synchronous replication*. Synchronous replication is discussed in Chapter 4, "*Synchronous Replication*."

One way to implement synchronous replication is to use distributed transactions to start a single transaction that spans both systems. Then updates to all data item instances are guaranteed, or else none will be made (the transaction aborts).

Distributed transactions may work well in campus environments but are inappropriate for geographically dispersed networks. This is because of the many channel propagation delays caused by the communication channel round trips that are required for a transaction – two per update (a read followed by a write) plus two for the prepare and commit messages. Recognizing that round-trip signal propagation

Breaking the Availability Barrier

time between the U.S. coasts is about 50 msec. and that it can be 250 msec. between London and Sydney, it is clear that network transactions can add seconds to an application's transaction response time. This is called *application latency*.

Geographically dispersed systems are a requirement for disaster recovery. For these systems, data replication with coordinated commits offers a satisfactory alternative[31] as described in Chapter 4. With this technique, the data replicator will start independent transactions on each system. Updates are replicated asynchronously, but the commits of these independent transactions are coordinated by the data replicator. Neither transaction is committed unless both are guaranteed to commit.

By using synchronous replication, both systems may participate as equals in the transaction. If one system goes down, the users on the other system are unaffected and can continue to process transactions while the downed users are switched over to the surviving system.

Recovery Time

Recovery time, and hence RTO, can be vastly different for these recovery strategies, as shown in Figure 6-3.

No Replication

No Replication, Periodic Backup Only

If data replication is not used, the current database must be restored at the backup site. This involves first retrieving the appropriate backup tapes and then getting them to the backup site. The latest snapshot (or even an entire backup) must be loaded.

---

[31] Holenstein, B. D.; et al.; *"Collision Avoidance in Database Replication Systems,"* United States Patent Application 20020133507; Sept. 19, 2002.

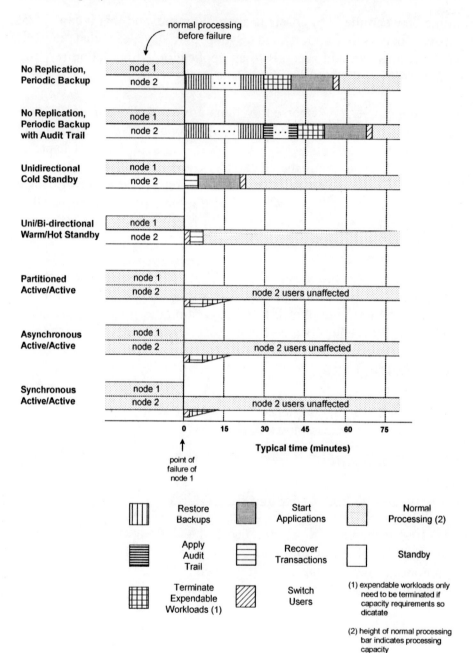

**Typical Recovery Times
Figure 6-3**

At this point, non-critical workload can be shed; and the applications to be recovered are started. Finally, users can be switched from the off-line system to the backup system; and normal activity can continue.

The reconstruction of the database can take several hours to several days, depending upon its size. The time required for the other functions of expendable workload shedding, application startup, and user transfer usually pales in comparison to database recovery time. In fact, these functions can be done during database recovery.

No Replication, Periodic Backup with Audit Trail

A major problem with the simple tape backup procedure discussed above is that there can be hours or days of data updates that are lost from the time of the last backup or snapshot to the time of the system failure. This results in a very large RPO. The RPO can be significantly reduced if provisions have been made for audit trail tapes which record all updates made to the database since the last backup or snapshot.

The use of audit trails not only results in a lower RPO, but also can bring the database to a consistent state by completing incomplete transactions or by purging aborted or incomplete transactions.

However, the use of audit trail tapes has the effect of increasing the recovery time, or RTO. Not only must the database be restored, but now the audit trail tapes must be re-played to bring the database more up-to-date.

Unidirectional Replication

Unidirectional Replication – Cold Standby

If data is being replicated to a cold standby, and if the standby must take over processing, then the applications must be started and the users switched over before service can be restored. Application startup generally requires many minutes before users can reinitiate their sessions and continue their activities. In addition, incomplete transactions must be purged. This may take a few more minutes.

Unidirectional Replication – Warm or Hot Standby

If data is being replicated to a warm or hot standby, then all that is required is for incomplete transactions to be recovered or aborted and for the users to be switched over. Also, if the backup is a warm standby, files must be reopened for full access. User switching can be accomplished in seconds, especially if switchover is automatic upon detection of a non-operating primary. File reopening is usually a matter of seconds for smaller applications but may be minutes for larger applications. Moreover, it may take several minutes to restore all incomplete transactions.

Bi-Directional Replication

Hot Standby

So far as recovery time is concerned, a hot standby bi-directional replication system recovers in the same recovery time as the unidirectional hot standby system described above.

Partitioned Active/Active Replication

The recovery from a system failure when using bi-directional replication is similar to that of unidirectional replication to a hot standby. The downed users simply need to be switched over to the surviving system, and transactions need to be recovered. Transaction recovery is usually accomplished by each application (and perhaps ultimately by each user) checking that the last transaction was accepted. If the transaction was not accepted, it is resubmitted. The other users continue their activities on their application partitions. However, the partitioned activities of the downed users are terminated until recovery is complete.

The surviving system becomes the owner of both database partitions. In this case, it may be desirable to shed some load by stopping non-critical applications in order to provide enough capacity for all users.

Asynchronous Active/Active Replication

If asynchronous active/active processing is being used, there is no impact on the users connected to the surviving system. They continue to be provided with all services. The loss of a system is only the loss of capacity.

The downed users must be switched over to the surviving system. Following recovery of transactions that had been in process on the failed system, and perhaps the shedding of some non-critical applications, full service can be restored.

Synchronous Active/Active Replication

Likewise, if synchronous active/active replication is being used, there is no impact on the users connected to the surviving system. They continue to be provided with all services. The loss of a system is only the loss of capacity.

After the downed users have been switched over to the surviving system and perhaps after shedding some non-critical applications, full service can be restored. No transactions need to be recovered since any transactions being processed by the failed system at the time of failure will be aborted.

Data Loss

RPO is the goal for the maximum amount of data that may be lost due to a failure. The amount of data that may be lost varies dramatically with these techniques.

Dr. Bill Highleyman, Paul J. Holenstein, and Dr. Bruce Holenstein

No Replication

If no audit trail tapes are used, all data since the last backup is lost. This can be hours or even days worth of data.

If audit trail tapes are used, data loss depends on whether some of these tapes were destroyed in a disaster. Data loss can range from seconds to hours.

Asynchronous Replication

Asynchronous data replication may be used with either unidirectional or bi-directional replication. That is, the source application does not wait for the data to be safely stored and/or applied at the target system. The interval between the time that an update is applied to the source database and the time that it is applied to the target database is known as replication latency. Replication latency is the time that replicated data is in the replication pipeline.

In the event of an outage, data in the replication pipeline may be lost. This typically represents one to thirty seconds of database updates, providing that updates are sent to the target system as soon as they have been applied at the source system. If transactions are held at the source system until they have been committed before sending them to the target system, then replication latency and therefore the amount of lost data can be significantly higher. This is especially true if long-lasting transactions (such as batch transactions) hold up a completed transaction and prevent it from being sent over the replication channel.

Synchronous Replication

Data collisions can be avoided by the use of synchronous replication. The source application does not commit a transaction until it is assured that the transaction will complete on the other system. If it is unable to commit, then the transaction is aborted at the

Breaking the Availability Barrier

other system. Therefore, there is no data loss (zero RPO) under normal operations if synchronous replication is used.

All synchronous replication methods have a very small but very critical uncertainty window. For instance, with networked transactions that use a two-phase commit protocol, if the network should fail between the receipt of the phase 1 response (ready to commit) and the receipt of the phase 2 response (commit), the source system does not know whether the target system received the commit command. Conversely, if the target system did not receive the commit command, then it does not know what to do with the transaction – commit it or abort it. This is often referred to as a *hung* transaction.

A similar window of uncertainty exists for coordinated commits. However, in either case, data changes are not lost. They are simply held in a locked state at the target system. This gives an opportunity to query the source system to determine whether the transaction actually completed. Assumed here, of course, is that the source system is available and accessible. Otherwise, the application or the user will have to determine whether or not to commit or abort the transaction at the target system. This situation will self-correct via the inherent recovery facilities of the replication engine should the source system once again become available.

Recovery Strategies

If either the primary or backup system should fail, or if a network fault should isolate a system, then the failed system's database will become rapidly out-of-date as processing activity continues at the surviving system. When the failed system is returned to service, its database must be brought to the current state.

If magnetic tape is being used to back up a system, then the database recovery of the failed system is straightforward. The latest tape copy of the database and any change tapes are simply loaded onto the recovered system, and operation proceeds as usual.

The recovery of a downed or isolated system when data replication is being used is somewhat different and is, in some cases, more difficult.

Unidirectional Replication

Asynchronous Replication

If data is being replicated to a cold, warm, or hot standby, and if data replication is asynchronous, then no special action need be taken. During the period of failure, the queue of transactions to be replicated simply grows at the source system.

When the failed system (or the interconnecting network, whichever caused the failure) is returned to service, asynchronous replication continues; and the queue to the downed system is drained. When the queue size falls below a level considered to be normal, the failed system can be considered to be recovered.

Synchronous Replication

If synchronous replication is being used, then the active system (which now may be the original backup system if the primary system failed) must switch to asynchronous replication and queue new transactions for later replication. When the failed system is returned to service, the queue must be drained. The saved transactions are replicated to the system being recovered.

The active system continues to function during recovery, using asynchronous replication to the recovering system. It will do so until the replication latency falls to the point that it will not seriously affect the response times of subsequent synchronously replicated transactions (i.e., the application latency is within acceptable bounds). At this point, synchronous replication is resumed; and synchronous replication messages are queued behind the remaining asynchronous replication messages.

Active/Active Replication

System Failure

If both systems are processing transactions in an active/active configuration, and one system fails, then the affected users may be switched to the surviving system and full functionality continued (within the capacity, of course, of the surviving system). Operation and recovery are as described above with respect to unidirectional replication, with the surviving system queuing transactions that must be posted to the failed system.

When the failed system is recovered, the transaction queue is drained while normal processing continues at the active system. Users may then be reconnected to the recovered system.

If asynchronous replication is being used, users may be switched when the queue size is small enough to reduce the chance of data collisions to an acceptable level.

If synchronous replication is being used, then the queue is drained to an acceptable level. At this point, normal processing can proceed with synchronous replication, as described above with respect to unidirectional replication.

Network Failure

If the network connecting the two systems should fail during bi-directional replication, then both systems are capable of continuing independent processing. However, in the general case, this can lead to extensive data collisions if the network outage is lengthy. There are several options for continuing operation in this case:

a) If the application can be partitioned so that any data item can be modified only by users at one system, then do so (partitioning may already be in effect). Each system can then continue to function independently and to queue its transactions to the other system for later replication when the network becomes operational. Recovery procedures for asynchronous and synchronous replication are described above.

b) If the application cannot be partitioned, and if data collisions are not acceptable, then all users can be switched to one of the systems. The isolated system is, in effect, down and is recovered as described above when the network become available.

c) If data collisions can be tolerated, then each system can continue to function and to queue its transactions for later delivery, with recovery proceeding as described previously. If replication is asynchronous, no change in operation occurs. Rather, the replication latency time is now the network down time. If synchronous replication is being used, the systems must switch to asynchronous replication until recovery, as described above. In either case, data collisions occurring during the outage must be detected and resolved.

Comparison Summary

Recovery times (RTO), data loss (RPO), and the possibility of data collisions for these various techniques are summarized in Figure 6-4. As can be seen, if replication is not used, both RTO and RPO can be very high – it may take hours to days to recover; and hours to days of data may potentially be lost.

If replication of any kind is used, RTO is reduced dramatically to minutes, seconds, or even to near-zero if active/active replication is used (depending upon the time that it takes to switch users on the failed system to the active system). Likewise, the amount of data that may be lost, as measured by RPO, may be as little as those changes which occurred in the few seconds before failure or, in the case of synchronous replication, may even be reduced to zero.

Even when replication is used, the RPO and RTO that can be achieved is very dependent upon the replication technology that is used. This is illustrated in a general way in Figure 6-5.

As shown in Figure 6-5, synchronous replication achieves a zero RPO. That is, no data is lost when a system fails. Any transaction in process at the time of the failure is aborted (as described earlier, there is a very small uncertainty window which leads to locked data but not to lost data). Furthermore, if the application is configured as an active/active application, RTO is also zero for all users but those on the failed system. Even the users on the failed system can be quickly restored to service by switching them to a surviving system.

The amount of data which may be lost with asynchronous replication is a direct function of the replication latency of the replication channel. If replication is substantially process-to-process (i.e., changes are extracted directly from the source database and applied directly to the target database with no intermediate queuing), replication latency can be quite small (sub-second) and RPO can be driven fairly close to zero. To the extent that there are queuing points in the replicator or that there are intermediate disk storage points, replication latency will be longer and RPO higher. Typical queuing points occur at the communication channel as well as at processes that are reading source data or are updating target data. Process queuing can be reduced by multithreading but perhaps at the expense of referential integrity as described in Chapter 10, "*Referential Integrity.*" Communication channel queuing can be reduced by using smaller block sizes at the expense of communication channel utilization efficiency.

Recovery time, which directly affects RTO, is also a function of several factors. If the application processes in the backup system are inactive, they have to be activated; this can take several minutes. However, if the application is configured as an active/active application, then the only requirement is for the users on the failed system to be switched to the surviving system. The RTO is zero for all other users.

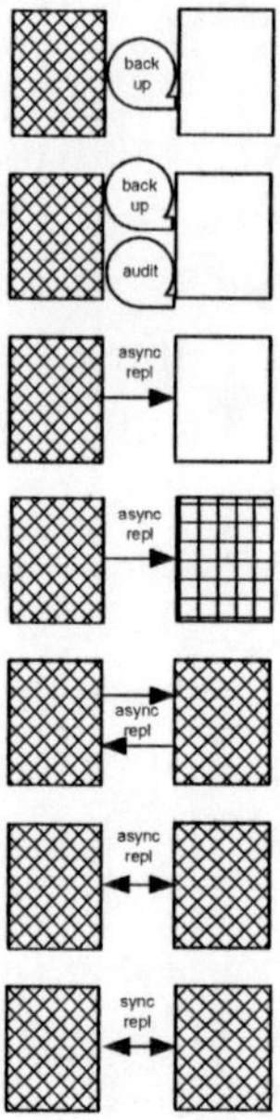

Type	Recovery Time (RTO)	Data Loss (RPO)	Data Collisions
No Replication, Periodic Backup	hours to days	hours to days	no
No Replication, with Audit Trail	hours to days	seconds to hours	no
Unidirectional Cold Standby	minutes	< 1 second	no
Uni/Bi-directional Warm/Hot Standby	seconds	< 1 second	no
Partitioned Active/Active	seconds	< 1 second	no
Asynchronous Active/Active	none	< 1 second	yes
Synchronous Active/Active	none	none	no

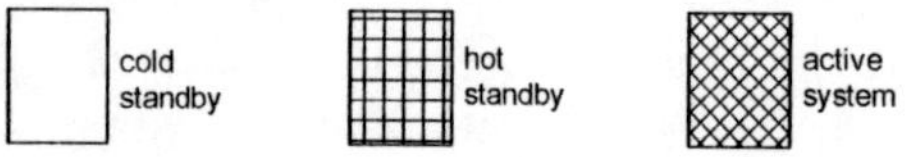

Replication Method Comparison
Figure 6-4

Furthermore, if the replication scope is an update rather than a transaction, or if the replication engine does not preserve the source transaction boundaries, then the database may not be in a consistent state. It will have to be cleaned up by determining incomplete transactions and backing them out, which could take a significant amount of time.

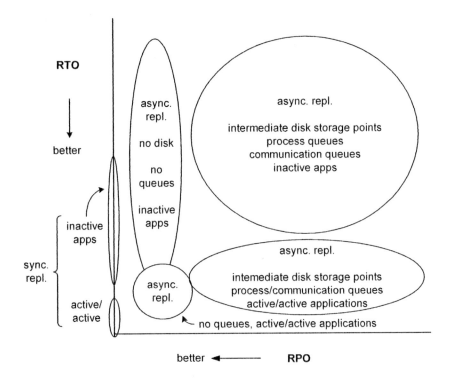

**The Impact of Replication Technology on RPO and RTO
Figure 6-5**

The above recovery time comments apply to asynchronous replication and to a lesser extent to synchronous replication. With synchronous replication, the database will not have to be cleaned up because all changes are within the scope of a transaction. However, synchronous replication can be used with a cold standby, in which case recovery time necessarily includes the activation of the application processes in the standby system.

Thus, as summarized in Figure 6-5, the actual RPO and RTO that can be achieved are not just a matter of synchronous versus asynchronous replication but are strongly dependent upon the specific technology used to implement the replication facility.

Rule 28: *RPO and RTO are both a function of the data replication technology used to maintain databases in synchronism.*

Multi-Node Applications

Any of these data replication techniques can be extended to multi-node applications to provide multiple recovery sites or to provide backup at one site for multiple primary sites.

Moreover, bi-directional replication can be used to spread an application load over several nodes, as described in Chapter 2, "*System Splitting.*" One advantage of this is a significant increase in full capacity availability. Another advantage is that if a node fails, only a portion of system capacity is lost. Except for users at the failed node, substantially full service continues to be provided. Moreover, the downed users can be returned to service within seconds by switching them to surviving nodes. Their only penalty is a session loss which they may not even recognize as a system failure since session loss due to communication problems is often a frequent occurrence anyway.

Furthermore, the loss of two nodes-worth of capacity is so unlikely that it typically can be ignored.

One problem with multi-node bi-directional replication when using partitioned applications is that database partitioning must be finer grained. Moreover, the transfer of ownership of a failed partition must be resolved. This is not a problem with active/active bi-directional replication because database partitioning is not required.

Recovery Decision Time

Should a disaster destroy a site, the site's inability to function is pretty obvious. There is not much need for a decision process to decide to activate the backup site.

However, replicated architectures also protect against normal system outages. Now when a failure occurs, a decision must be made whether to recover the failed system or to switch over to the backup system.

This can be a high-stress process. Remember Bartlett's Law (Rule 19) - *When things go wrong, people get stupider.* First, operational personnel must acknowledge that there is a problem. Then the problem must be diagnosed and consensus reached on the optimum recovery action. Often, management must be consulted to approve any drastic actions such as a switchover and the termination of other applications. This process can take an hour or two, especially if it happens during off-hours.

Thus, for non-disastrous system outages, the time lines in Figure 6-3 should be preceded by an extended decision-making time. This directly impacts the non-replication and unidirectional replication methods. It only affects the downed partition for partitioned bi-directional replication.

Active/active bi-directional replication does not suffer a decision time since these applications keep on running in the presence of a node outage. The only decision necessary is whether to switch the users at the failed site to the surviving site.

Summary

In the early days of computing, only tape was available for backups. In those days, the batch nature of computing made long RPOs and RTOs acceptable. However, as systems came online and as real-time became a reality, RPO and RTO became more important. This importance only increased as systems became more-mission critical to enterprises.

Replication technology came about to satisfy the needs of these real-time, mission-critical systems. Early replication facilities reduced RPOs and RTOs from days or hours to minutes. As pressure increased to further improve these objectives, replication technology improved to reduce replication latency and shorten recovery times. The use of higher speed communication channels and multithreaded replicators and the elimination of intermediate disk storage points have all contributed to faster replication and less loss of data. The tolerance of replication facilities to provide active applications at the backup have greatly speeded recovery time.

Recent solutions to critical bi-directional replication problems such as the ping-ponging of updates and the detection and resolution of data collisions have allowed active/active systems to become a reality. Active/active applications allow all of the data processing power in a network to be utilized. Now with the inherent increase in computing system power which compensates for the additional overhead inherent with synchronous replication, efficient active/active systems can be built using synchronous replication. These systems will not only reduce RPO to zero, but they will also virtually eliminate RTO as a consideration. Recent advances in more efficient transaction coordination over wide area networks (coordinated commits) have reduced the performance impact of synchronous replication even further.

As a result, the technology to build extremely highly available and disaster-tolerant systems at little additional cost, with no data loss due to a network or node failure, and with no service loss to users except briefly at a failed node is here today.

What's Next

We have now concluded our discussion of ways to split systems into independent cooperating nodes to achieve significantly higher availabilities. We have explored the various methods that can be used to keep the database copies which are distributed across the network in synchronization. We have considered the impact of hardware faults, software faults, failover faults, and recovery times on system availability, and have described some important measures of distributed system effectiveness.

In the next Chapter, "*The Ultimate Architecture*," we put all of this together to suggest architectures that can provide significantly increased availability at little or no additional cost using what we have learned so far.

Chapter 7 - The Ultimate Architecture

In this chapter, we apply the concepts that we have generated in our previous six chapters to suggest an ultimate architecture that can extend the four 9s availability of today's systems to six, seven, or even eight 9s at little additional cost.

In our previous chapters, we looked at several aspects of system availability:

- We analyzed the availability of a system in terms of the reliability of its component subsystems and its failure modes.

- We pointed out the reliability advantages that can be gained by splitting a system into several smaller cooperating but independent systems.

- We explored various methods for keeping these independent systems synchronized with each other.

- We considered the implications of system outages that are due to software failures or human errors and that are corrected by recovery rather than repair.

- We pointed out the compromises between data replication techniques and the twin objectives of recovery time and data loss following a failure.

Before we look at ultra-high availability architectures, let us review what we have learned so far.

Dr. Bill Highleyman, Paul J. Holenstein, and Dr. Bruce Holenstein

An Availability Review

In Chapter 1 of this series, we analyzed the availability of redundant systems comprising a number of identical subsystems. Should enough subsystems fail so that an outage occurs, the system is restored to service as soon as the requisite number of subsystems is repaired. We found that the system availability A can be expressed as

$$A = \frac{MTBF}{MTBF + MTR} = 1 - F \qquad (7\text{-}1a)$$

where

$$F \approx \frac{MTR}{MTBF} \approx f(1-a)^{s+1} \approx f\left(\frac{mtr}{mtbf}\right)^{s+1} \qquad (7\text{-}1b)$$

and where

- A is the system availability.
- F is the system probability of failure.
- $MTBF$ is the system mean time before failure.
- MTR is the system mean time to restore (repair plus recovery).
- s is the number of spare subsystems provided.
- f is the number of failure modes, or the number of ways in which $s+1$ subsystems can fail such that a system outage is caused.
- a is the subsystem availability.
- $mtbf$ is the subsystem mean time before failure.
- mtr is the subsystem mean time to repair.

For systems configured with a single spare, if any failure of two subsystems can cause a system outage, then the number of failure modes is

$$f = \frac{n(n-1)}{2} \qquad (7\text{-}1c)$$

where

Breaking the Availability Barrier

n is the number of processors in the system.

We also showed that

$$\text{MTR} = \frac{\text{mtr}}{\text{s}+1} \tag{7-2}$$

$$\text{MTBF} \approx \frac{\text{mtbf}}{f(\text{s}+1)} \left(\frac{\text{mtbf}}{\text{mtr}}\right)^{\text{s}} \tag{7-3}$$

We pointed out that the number of failure modes in a single-spared system is very sensitive to the allocation of processes to processors, and poor allocation can reduce system reliability by more than an order of magnitude.

In Chapter 2, we looked at the dramatic improvements in availability obtained by replicating systems - an approach, however, that is very expensive. We extended the replication concept to the more economical approach of splitting a system into k smaller independent but cooperating systems. We found that such a network of systems is at least k times more reliable than a single system in terms of providing 100% capacity. Moreover, we found that in the event of a system outage, only $1/k$ of the total processing capacity is lost rather than all of it. Furthermore, the chance of losing more than $1/k$ of the system capacity is almost never.

Of course, these k independent systems must keep their databases synchronized. In Chapters 3 and 4, we looked at techniques for doing this and evaluated the transaction performance of two key methods for providing exact database synchronization. The two methods are dual writes within a single transaction (network transactions) and coordinated commits. The latter method involves starting independent transactions on each system, replicating data updates asynchronously, and then coordinating the commits at each system.

The systems studied up to this point were repairable systems. That is, in the event of an outage caused by $s+1$ subsystem failures, the repair of one failed subsystem allows the system to be returned to service. In Chapter 5, we considered the reality of today's fault-tolerant systems – most system outages are caused by software faults or human errors. As a consequence, a system usually does not have to be repaired following an outage; it has to be recovered. We also considered the impact of failover faults on system availability. We quantitatively demonstrated the importance of short recovery times in minimizing the impact of failover faults and for improving system availability in general.

In Chapter 6, we argued that the data replication method of choice depended upon the application's tolerance to recovery time and lost data. We showed that these were independent considerations and were defined by the data replication technique chosen.

In this chapter, we put together the concepts of our earlier chapters to suggest an ultimate architecture that can potentially increase system reliability by several orders of magnitude at perhaps little additional cost. We start with a standard single-spared fault-tolerant system as a base line. We then consider a variety of high-availability options leading us to the suggested ultimate architecture. As we progress, we will move from today's infrastructure into that which we hope will be available tomorrow.

The Strawman System

We start with a single 7x24 fault-tolerant system that we will re-architect to improve its availability (Figure 7-1). The system has the following parameters:

Number of processors	16
Mean time before failure	5 years
Mean time to restore (recovery)	4 hours
Availability	(5 years) / (5 years + 4 hours) = .9999

Note that the five-year MTBF assumption includes outages from all causes, including operational errors, hardware failures, software faults, application bugs, and environmental problems.

Splitting Into Independent Systems

Our first attempt at improving availability is to replace the single system of Figure 7-1 with a network of smaller systems, each being fully independent and in total providing the same capacity as the single system. For purposes of illustration, we will split the 16-processor strawman system of Figure 7-1 into four 4-processor systems (Figure 7-2), each with a mirrored copy of the full database.

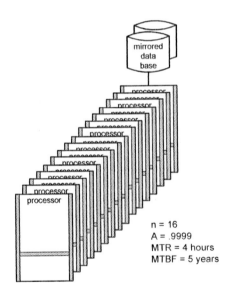

**Strawman Single System
Figure 7-1**

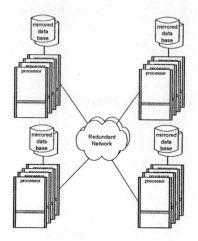

**Split System
Figure 7-2**

We assume that users at any system can update any data item in the database (after all, that is what they can do in the strawman system). An update made at one system must then be propagated to the other systems in the network via some means such as data replication. This is what we have called an *active/active* application.

We explained in Chapter 2 that splitting a system into k nodes reduces its probability of failure (i.e., increases its reliability) by a factor of

$$k\frac{n-1}{n-k} > k \qquad (7\text{-}4)$$

where

- n is the number of processors in the original single system.
- k is the number of nodes into which the original system is split.

Note that this factor is always greater than k. Splitting a system into k nodes increases its reliability by at least k.

Breaking the Availability Barrier

Let us define an outage as the loss of just one of the k nodes. Thus, if the system fails, we lose just $1/k$ of its capacity, not 100% as we will with a single system.

In the case shown in Figure 7-2, we have split the single system into four nodes $(k=4)$. Therefore, from Expression (4-4), this split system will be five times more reliable (4x15/12) than the single system and will provide an availability of .99998 rather than .9999. This means that the system MTBF has increased from 5 years to 25 years. As an added plus, when it does fail, it still will provide 75% of its capacity (the single system will provide no capacity).

Even more striking is the split system's tenaciousness to provide at least 75% capacity. Note that it will take the failure of two nodes to reduce the system capacity to less than 75%. There are six ways that the four-node system of Figure 7-2 can lose two nodes. Therefore, from Equations (7-1b) and (7-1c), the probability of a two-node failure, F, is $6(1-.99998)^2$, which is more than nine 9s. From Equation (7-1a), we also note that

$$\text{MTBF} \approx \frac{\text{MTR}}{F} \qquad (7\text{-}5)$$

Using our MTR assumption of four hours, the average time before losing 75% capacity (i.e., a two-node failure) is over 1,900 centuries!

Another advantage of the split-system architecture is that the application survives even if the network fails, though the system continues with disconnected independent nodes.

However, this architecture comes with one big problem – data collisions. There is nothing to prevent two users at two different systems from updating the same data item at the same time, thus putting the database in an inconsistent state. The detection of data collisions can impose significant overhead on the system. Even worse, the resolution of data collisions often is a manual process.

Dr. Bill Highleyman, Paul J. Holenstein, and Dr. Bruce Holenstein

In Chapter 4, we showed that even in reasonably sized systems, collision rates can easily exceed 1,000 collisions per hour. If they must be resolved manually, this situation is clearly untenable. Some applications such as security trading systems cannot tolerate data collisions at all.

One solution to avoid collisions is to use synchronous replication, as described in Chapter 4. However, the complexity of the transaction grows as more nodes are added to handle additional traffic. For eight nodes, each transaction will have to make 8 times the number of updates associated with each transaction. This will certainly be undesirable using dual writes (i.e., all updates are done under a single transaction). Large numbers of updates in a transaction clearly call for coordinated commits using asynchronous data replication, as is pointed out in Chapter 4. However, even in this case, the network traffic grows as k^2 (each new node adds another batch of transactions, and each transaction is now longer). Therefore, the system is not scalable.

Another problem with such an architecture is cost. Splitting processors among the nodes is cost-efficient. However, reproducing the database at each node can be extremely costly since in many large single systems, the database represents 70% to 90% of the system cost.

No wonder, as Jim Gray[32] points out, we don't see large active/active applications being deployed today.

System Splitting with Dual Databases

As pointed out above, the architecture shown in Figure 7-2 has three severe problems: data collisions, scalability, and cost.

---

[32] Gray, J.; et al.; "*The Dangers of Replication and a Solution*," ACM SIGMOD Record (Proceedings of the 1996 ACM SIGMOD International Conference on Management of Data), Volume 25, Issue 2; June, 1996.

All of these shortcomings can be substantially improved by recognizing that the mirrored database does not have to be replicated across all systems. It is sufficient to have only two mirrored copies of the database in the network so that the database will still be accessible in the event of a node failure. Figure 7-3 shows a configuration in which the database is split into k partitions a, b, c, and d (four in our case) and in which each partition has one mirror a', b', c', and d' on another node.

Now we need only to pay for two databases, regardless of the size of the network. Furthermore, since each update need only be made to two copies of the database, network traffic grows only as the transaction rate grows. As a result, the system is scalable.

Synchronous replication now becomes a real option to keep the databases synchronized to avoid data collisions. The number of database actions required only doubles regardless of the number of nodes (in our previous discussion, it was proportional to the number of nodes), and the network traffic increases proportionally to the transaction rate rather than to the square of the number of nodes. This is a scalable solution.

So far as the method for synchronous replication is concerned, we showed in Chapter 4 that dual writes to both partitions as part of a common transaction (a network transaction) is applicable for short transactions within co-located nodes such as campus configurations. Coordinated commits using data replication is appropriate for wide-area configurations or large transactions. This is discussed further below.

Do We Need to Replicate a Mirrored Database?

A mirrored database is already redundant. Why do we need two of them? After all, we were satisfied with a single mirrored pair in our strawman 16-processor system.

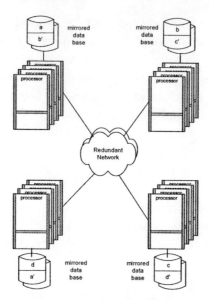

**Split System, Dual Mirrors
Figure 7-3**

More specifically, a typical single disk today has a mean time before failure of about 500,000 hours. Let us de-rate this to 100,000 hours to account for environmental and other degrading factors. Furthermore, let us assume a leisurely 24-hour repair time. The mirrored disk system has one spare ($s=1$) and one failure mode, which is the failure of both disks ($f=1$). From Equation (7-3), the MTBF of a mirrored disk pair is nearly 500 centuries! Its availability is over eight 9s.

Our single fault-tolerant system was assumed to have an MTBF of five years. Our split system is five times more reliable and therefore has an MTBF of 25 years. The mirrored disk pair is orders of magnitude more reliable. Therefore, the system only needs a single mirrored database.

However, we do not want to simply connect our mirrored pair to one of the nodes because the failure of that particular node will take down the entire system. We have other options as follows.

Breaking the Availability Barrier

Option 1: Split Mirrors

We can split our disk mirror between two nodes, as shown in Figure 7-4a. Now the failure of any one node does not take down the system. Such a failure's only impact is to lose $1/k$ of the system capacity (25% every 25 years on the average, in this case). Losing another chunk of $1/k$ capacity or, even worse, both database nodes, will happen almost never. We have all of the advantages of system splitting (almost five 9s availability) at virtually no extra cost (the same number of processors and disks as the single system) and with a small performance penalty caused by the requirement for synchronous replication.

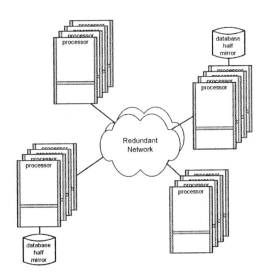

Split Mirrors
Figure 7-4a

Option 2: Network Storage

Alternatively, the split mirrors of Figure 7-4a can be made to be independent of any processor, as shown in Figure 7-4b. This is the promise of network storage. The disk subsystem will connect independently to the network rather than to a processing node.

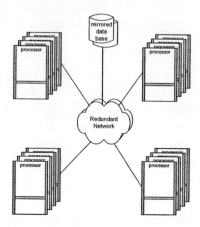

Mirrored Network Storage
Figure 7-4b

There are several advantages to this configuration. The loss of any one node in the network will not cause the loss of both a processing node and a database node. Furthermore, the system will survive multiple failures of processing nodes, albeit with reduced capacity. Finally, a processing node can be taken down for maintenance or update without compromising the availability of the database.

However, the failure of both database mirrors will cause a system failure, though we have argued that the probability of this happening is orders of magnitude less likely than the loss of a processing node. Of course, this does not consider a disaster that takes out the database system or the network connecting it to the processing nodes. If disaster tolerance is a requirement, then the architecture of Figure 7-4a is appropriate and uses data replication with coordinated commits to keep the database in synchronism.

At some time in the future, if geographically distributed network storage should become available, then even the configuration of Figure 7-4b can provide disaster tolerance. Geographically distributed network storage must await distributed disk managers with distributed lock management.

The Ultimate Architecture

The configurations described for single mirrors work for small disk farms. Our examples have been based on a database comprising a single pair of mirrored disks.

But what about large disk farms? A disk system with d mirrored pairs will have an MTBF of $500/d$ centuries since there are d ways in which it can fail. A large disk farm with 1,000 mirrored pairs will have an MTBF of 50 years, comparable to the processor network. Larger disk farms will degrade the system availability further.

The solution? Build the disk system with a sparing level of two ($s=2$ in Equations (7-1) through (7-3)). Using our previous parameters, such a triply-redundant disk system will have an MTBF (using Equation (7-3) with $f=3$, $s=2$) much longer than the earth is expected to last. Even for large disk farms, the disk system will have MTBFs measured in earth life times and can be ignored so far as availability is concerned.

Expanding on our architectures of Figures 7-4a and 7-4b, we now have the architectures shown in Figure 7-5. The system will have to lose three disk subsystems to create an outage. As calculated earlier, if it loses one processing node, it loses 25% of its capacity, which is expected every 25 years, on the average. If it loses two processing nodes, it loses 50% of its capacity, which is expected every 1,900 centuries, on the average. These results are summarized in Table 7-1.

How can we implement triple redundancy? One option is to provide each disk volume with two mirrors, as shown in Figure 7-5. This represents additional cost, but it is just 50% above the cost of our mirrored system disk cost.

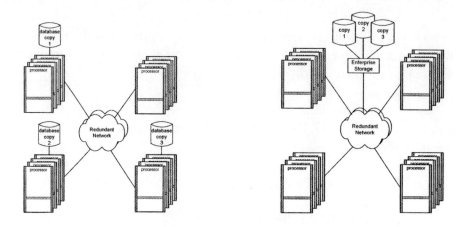

Split System with Triply Redundant Database
Figure 7-5

		Capacity		
		100%	75%	50%
Single System - 16 processors	Availability	.9999	---	---
	MTBF (years)	5		
Split System - four 4-processor nodes	Availability	.99998	9 9s	21 9s
	MTBF (years)	25	190,000	Almost always

Comparative Availability of Split System
Table 7-1

Even so, this added cost can be close to a 50% penalty on the entire system cost if the disk system represents 70% to 90% of the single system cost. The answer to the cost hurdle may be RAID. RAID arrays are redundant arrays of independent disks or redundant

Breaking the Availability Barrier

arrays of inexpensive disks, depending upon your outlook. Basically, RAID involves the striping of data across several disks with striped parity blocks that allow the reconstruction of lost data due to one or more failed disks. The popular RAID 5 provides protection against a single disk failure by providing a single parity stripe. If *N* disks are needed to store the data, *N+1* disks are needed for RAID 5.

The new RAID 6 configuration[33] provides dual parity striping over *N+2* disks and can survive dual disk failures. This is the triple redundancy for which we are looking.

These RAID disk configurations are so reliable that inexpensive disks can be used. So, concentrating on RAID as being random arrays of *inexpensive* disks, we have a system that

- has five to nine 9s availability, depending upon capacity requirements.

- is highly scalable.

- is no more costly than an equivalent single system that uses today's disk technology (but be careful of additional software licensing and operational costs).

- provides active/active application support with no data collisions.

This, we submit, is the ultimate availability architecture.

Performance Impact of Synchronous Replication

Split system architectures which must use synchronous replication because they cannot tolerate data collisions suffer a performance hit because of the requirement to keep remote databases in synchronism. Cross-country, round-trip communication channel delays can be 50

---

[33] Advanced Computer and Network Corporation, "*RAID 6,*" www.acnc.com.

msec. (at half the speed of light); and an application must wait on these delays before it can commit a transaction.

In Chapter 4, we showed that dual writes under a single transaction required two round-trip delays for each update plus two more for the commit. For campus environments with channel delays measured as a few milliseconds, this method may work well. However, replicating over long distances or replicating long transactions can easily add seconds to modestly sized transaction.

For larger transactions, for long distances, or for several database copies, coordinated commits or the equivalent will perform better. Using this technique, updates generated by the source transaction are sent to the target databases via asynchronous data replication. The transactions started at each target are coordinated with the source transaction, and a system-wide commit is executed only if all targets concur that their commits will be successful. We showed in Chapter 4 that this entails two communication channel delays plus a data replication latency. Typically, this will add a small fraction of a second to a transaction. Note that this does not affect a node's throughput. It simply means that more servers in a server class will be required to handle the longer running transactions.

The good news is that if you are used to one-second response times and are upgrading to higher speed processors, you probably won't notice the difference in performance under coordinated commits because the synchronous replication delay will be compensated for by the higher speed of the new processors.

Here Come Local Clusters

For campus environments, high-speed communication fabrics for local environments are becoming a reality. These fabrics have tremendous capacity and sub-millisecond latency times. A good example is HP's NonStop ServerNet.

Breaking the Availability Barrier

Return for a moment to our 16-processor system split into four 4-processor nodes. Given the capabilities of ServerNet Clusters, for instance, our 16-processor system is child's play. Today's ServerNet Clusters allow up to 24 independent multiprocessor nodes to cooperate over a very high-speed redundant network, and this limit will be further relaxed in the future. Coupled with HP's Application Clustering Services (ACS), application domains may easily span multiple nodes in such a cluster.

Consider a twenty-node system. The loss of one node will cause the loss of just 5% of the system capacity, and full service will be provided to all users once the users on the failed node are switched over to surviving nodes. This is hardly an outage at all. The loss of 10% capacity (a two-node failure) is so unlikely that it may make the headlines of some future galactic virtual newspaper.

Database Replication – Enhancements Wanted

The real work to achieve the architectures discussed above is in the synchronous replication of the databases. Certainly today, this is achievable and has been described in earlier chapters.

However, there are several database management enhancements that we should put on our wish list, including:

- Support for distributed mirrors located at different nodes.

- Support for triply redundant distributed mirrors for very large systems.

- Network mirrored storage.

- Triply-redundant network storage (like RAID 6) for very large systems.

- Distributed network storage mirrors for disaster tolerance.

Conclusion

All of our work in the previous chapters has culminated in a fairly simple architecture that can double the nines of our current fault-tolerant systems. We have shown a path to designing systems which

- double the nines.

- provide active/active applications without data collisions.

- are scalable.

- are cheap.

- are achievable today.

We leave you with two final rules:

Rule 29: *You can have high availability, fast performance, or low cost. Pick any two.*

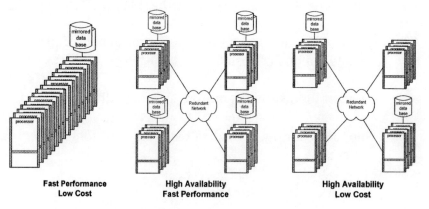

Availability, Performance, and Cost
Figure 7-6

Just remember –

Rule 30: *A system that is down has zero performance and its cost may be incalculable.*

Chapter 8 - The Rules of Availability

In Part 1 of this book, we have explored basic availability concepts and have applied them to various topics, including:

- software configuration
- system replication
- system splitting
- data replication
- repair and recovery time
- failover faults
- recovery point and recovery time objectives
- ultra-high availability architectures

As we worked through this material, we generated a series of availability rules. They encapsulate in a short-cut form most of what we discussed and are summarized below. Perusing these rules is an excellent review of all that we have discussed so far.

Rule 1: *If all subsystems must be operational, then the availability of the system is the product of the availabilities of the subsystems.*

Rule 2: *Providing a backup doubles the 9s.*

Rule 3: *System reliability is inversely proportional to the number of failure modes.*

Rule 4: *Organize processors into pairs, and allocate each process pair only to a processor pair.*

Rule 5: *If a system can withstand the failure of s subsystems, then the probability of failure of the system is the product of the probability of failures of (s+1) systems.*

Dr. Bill Highleyman, Paul J. Holenstein, and Dr. Bruce Holenstein

Rule 6: *System availability increases dramatically with increased sparing. Each additional level of sparing adds a subsystem's worth of 9s to the overall system availability.*

Rule 7: *For a single spare system, the system MTR is one-half the subsystem mtr.*

Rule 8: *For the case of a single spare, cutting subsystem mtr by a factor of k will reduce system MTR by a factor of k and increase the system MTBF by a factor of k, thus increasing system reliability by a factor of k^2.*

Rule 9: *If a system is split into k parts, the resulting system network will be more than k times as reliable as the original system and still will deliver (k-1)/k of the system capacity in the event of an outage.*

Rule 10: *If a system is split into k parts, the chance of losing more than 1/k of its capacity is many, many times less than the chance that the single system will lose all of its capacity.*

Rule 11: *Minimize data replication latency to minimize data loss following a node failure.*

Rule 12: *Database changes generally must be applied to the target database in natural flow order to prevent database corruption.*

Rule 13: *Follow natural flow order when replicating so as not to create artificial activity peaks at the target database.*

Rule 14: *Block the ping-ponging of data changes in a bi-directional replication environment to prevent database corruption.*

Rule 15: *Minimize replication latency to minimize data collisions.*

Rule 16: (Gray's Law) - *Waits under synchronous replication become data collisions under asynchronous replication.*

Breaking the Availability Barrier

Rule 17: *For synchronous replication, coordinated commits using data replication become more efficient relative to dual writes under a transaction manager as transactions become larger or as communication channel propagation time increases.*

Rule 18: *Redundant hardware systems have an availability of five to six 9s. Software and people reduce this to four 9s or less.*

Rule 19: (Bartlett's Law) - *When things go wrong, people get stupider.*

Rule 20: *Conduct periodic simulated failures to keep the operations staff trained and to ensure that recovery procedures are current.*

Rule 21: *System outages are predominantly caused by human and software errors.*

Rule 22: (Corollary to Rule 20) - *A system outage usually does not require a repair of any kind. Rather, it entails a recovery of the system.*

Rule 23: (Niehaus' Law) - *Change causes outages.*

Rule 24: *Following the failure of one subsystem, failover faults cause the system to behave as if it comprises n-1 remaining subsystems with decreased availability.*

Rule 25: *The possibility of failover faults erodes the availability advantages of system splitting (see Rule 9).*

Rule 26: (The Golden Rule) - *Design your systems for fast recovery to maximize availability, to reduce the effect of failover faults, and to take full advantage of system splitting.*

Rule 27: *Rapid recovery of a system outage is not simply a matter of command line entries. It is an entire business process.*

Dr. Bill Highleyman, Paul J. Holenstein, and Dr. Bruce Holenstein

Rule 28: *RPO and RTO are both a function of the data replication technology used to maintain databases in synchronism.*

Rule 29: *You can have high availability, fast performance, or low cost. Pick any two.*

Rule 30: *A system that is down has zero performance and its cost may be incalculable.*

Following is a preview of the rules formulated in the Advanced Topics chapters, comprising Part 2 of this book.

Rule 31: *Minimize lock latency to minimize synchronous replication deadlocks.*

Rule 32: *Lock latency deadlocks under synchronous replication become collisions under asynchronous replication.*

Rule 33: *Designating a master node for lock coordination can eliminate lock latency deadlocks when using synchronous replication.*

Rule 34: *Database changes must be applied to the target database in natural flow order to maintain referential integrity. (See Rule 12).*

Rule 35: *A serializing facility that will restore natural flow is required following all data replication threads and before the target database in order to guarantee that the database will remain consistent and uncorrupted.*

Part 2 - Advanced Topics

Chapter 9 - Data Conflict Rates

There are many reasons to distribute an application across multiple nodes:

- disaster tolerance.
- locality of data access.
- reduction of failure modes.
- service continuation during planned outages.
- guarantee of continued service in the presence of a node failure.

In the general distributed system case, any user at any node can update any data item in the database. This creates no problem if there is only one copy of the database in the system. However, a single database copy represents a single point of failure and does not satisfy any of the requirements enumerated above. It is therefore often desirable to have multiple copies of the database distributed in some way among the nodes. This may be done to improve fault tolerance, to provide for local data access, or to provide continued service during planned outages or node failures.

As modifications are made to one copy of the database, they must be replicated to the other copies so that all copies are kept in exact or near synchronism. The act of replication can, in many cases, create update conflicts in the database. Some conflicts are collisions, in which different copies of the same data item are updated simultaneously. Other conflicts are deadlocks and prevent the replication from continuing.

In previous chapters, we discussed in some detail data collisions, waits, and deadlocks and what to do about them. In Chapter 4, "*Synchronous Replication,*" we even quantified data collision rates with a magical formula. We now provide the analysis of the conditions under which these conflicts may happen and the expected rates at which they may occur.

Synchronous versus Asynchronous Replication

There are two general ways to replicate data – synchronously and asynchronously. With synchronous replication, all instances of a data item across the network are locked before any of them are modified; and then they are all modified and unlocked. That is, no data item modification is made permanently unless it is guaranteed that all copies of that data item can be so modified. In this way, all copies of the database are kept in exact synchronism. Modifications (i.e., inserts, updates, deletes) are often done as part of a transaction. This ensures that either all modifications within the scope of the transaction are made or that none are.[34]

Examples of synchronous replication methods include dual writes and coordinated commits:

- When using *dual writes* (more properly, *plural writes*), a transaction will first acquire locks on all copies across the network of all the data items which it wishes to modify. The transaction will then apply its modifications to those data items, following which it will release its locks. A global transaction manager will ensure that all of these modifications are applied, or that none are.

- Synchronous replication via *coordinated commits* depends upon the coordination of otherwise independent transactions across the network. As a transaction is applied at one node, the database modifications made by that transaction are asynchronously propagated across the network to the other database nodes. At each node, a new transaction is started; and

---

[34] This is the atomicity property of ACID transactions – atomic, consistent, independent, durable. See Gray, J.; et al.; Transaction Processing: Concepts and Techniques, Morgan Kaufman; 1993.

Breaking the Availability Barrier

the data items to be modified are locked within the scope of that transaction. However, none of these transaction updates are committed until the source node queries each remote node and ensures that each is ready to commit. At that point, it will commit its transaction and that of all remote copies.

With asynchronous replication, separate transactions are executed at each database copy. When a node receives a transaction from an application, the modifications called for by the transaction are applied to and committed at that node's database. The transaction is also passed on by that node to the other database nodes in the network. Each node will independently apply the transaction's modifications to its database.

Synchronous replication has the advantage that all database copies in the network are always in the same state, but it imposes a performance penalty on the application. This is because an application's transaction must wait for all database copies in the network to be modified before it can complete. This additional wait time is called *application latency*.

Asynchronous replication, on the other hand, imposes no performance penalty on the application because modifications to remote databases are made independently of the application. However, the databases are not kept tightly synchronized. Each transaction is applied to each database at a slightly different time. The time that it takes a source node to pass a transaction to a remote node is called *replication latency*. It is during the replication latency time that the databases are in different states.

The performance differences between asynchronous replication and synchronous replication have been considered in Chapters 3 and 4. However, there is another major difference between these two forms of replication and that is the nature of data conflicts that may occur. Data conflicts are the subject of this chapter.

Dr. Bill Highleyman, Paul J. Holenstein, and Dr. Bruce Holenstein

Deadlocks and Collisions

Whenever several application processes are updating the same database, there is a possibility of contention. That is, two or more processes may be trying to update the same data item at the same time. For instance, Process 1 may read data item A with the intent of modifying it. Before it can rewrite the modified value, Process 2 may also read data item A with the intention of modifying it. Process 1 will then rewrite its modified value, followed by Process 2. But the updated value rewritten by Process 2 will overwrite Process 1's value, thus losing it. (This is often referred to as the "lost update" problem.)

Such contention is usually resolved by locks. Before updating data item A, Process 1 will lock it so that no other process may access it. It will then read and update data item A and unlock it. When Process 2 attempts to read data item A, it will find that it is locked and will have to wait for it to be unlocked. When Process 1 releases the lock, then Process 2 may proceed with its update. (We ignore the effects of more complex locking protocols such as shared locks for simplicity sake.)

Locking creates the obvious problem of slowing down applications to some extent since they may have to wait on locks, especially in a busy system. This has the effect of reducing application concurrency but is a necessary evil. However, the nature of data contention leads to other problems - deadlocks and data collisions.

The data contention problems that we analyze below are the results of distributing an application across independent nodes and generally occur only if applications at more than one node can modify the same data item. Such contention can be avoided if the database can be partitioned in such a way that each data item can only be updated by one node. For instance, customers may be partitioned across nodes for update purposes. Alternatively, update privileges may be rotated among the nodes on a time schedule.

This analysis assumes that there is no partitioning of the database, that all nodes may modify all data items, and that updates are made randomly across the database.

Deadlocks

Mutual Waits

We discuss deadlocks first and start with those caused by *mutual waits*. Consider a single node system in which multiple application processes are actively applying transactions to the same database. Further consider the following scenario. Process 1 has locked data item A and also must update data item B. However, Process 2 has locked data item B and also must update data item A. Process 1 must wait on the lock on data item B held by Process 2, and Process 2 must wait on the lock on data item A held by Process 1. Neither process can proceed. We will call this a *mutual wait* deadlock. These types of deadlocks are typically resolved by one process timing out and releasing its locks. The second process can then proceed with its updates. The process which timed out can then retry its transaction, either automatically or by having the user resubmit it.

This same scenario can be extended to a multi-node configuration in which synchronous data replication is being used to keep the databases at each node in synchronism with each other. When a transaction starts at one node, it must acquire locks on all copies of a data item before it can permanently modify the data item copies. It is quite possible that two transactions, running on the same node or in different nodes, will deadlock as described above. One way to resolve such a deadlock is for one process to time out and abort its transaction, thus allowing the other transaction to proceed. The aborted transaction can then be retried.

When using asynchronous replication, there will be no inter-nodal deadlocks. However, the replication engine can deadlock with local applications or other replication engines running in the same node.

Dr. Bill Highleyman, Paul J. Holenstein, and Dr. Bruce Holenstein

Lock Latency

When locks are acquired across a network, there may be a time delay between the time that a lock is acquired at the source node and the time that it is acquired at a remote node. We call this *lock latency*.

It is quite possible that applications on two different nodes may acquire a lock on their local copy of the same data item within the lock latency time. In this case, when they attempt to lock the remote copy of that data item, each will wait on the lock acquired by the other application; and again a deadlock will occur.

Intelligent Locking Protocols

Mutual wait deadlocks can often be avoided by using an intelligent locking protocol, or ILP. For instance, if locks are always acquired in the same order (such as parent record first, then child records), then there will be no deadlocks.

An ILP will avoid deadlocks in a system with a single copy of the database. However, it will work for distributed database copies in only certain cases, as described later. The general use of ILPs is also discussed in more detail later.

The analysis that follows assumes the worst case in which locks are acquired in a random order and in which locking any data item can result in a deadlock. The results are thus conservative and represent an upper bound on deadlock rates.

Collisions

With asynchronous replication, there are no replication-induced deadlocks since modifications are applied independently to each database without concern for what is going on at other databases (though there could be local application process deadlocks which can be managed with an ILP or timeouts). However, because of replication latency, it is possible that two (or more) users at different nodes may modify the same data item at roughly the same time. Each of these nodes now has a different view of the value of this data item and will send its version to the other database nodes in the system. None of these views are correct, and the database is now in an inconsistent (and probably wrong) state. This is called a *collision*. Collisions represent a consequence of asynchronous active/active database replication. Collisions must be identified and resolved either automatically via business rules (which is not always possible) or manually, as described in more detail in Chapter 3.

Synchronous replication avoids collisions. In effect, *a wait for a lock in a synchronously replicated system translates to a data collision in an asynchronously replicated system.* This is Rule 16 given in Chapter 3. Both waits and collisions are caused by roughly simultaneous updates at two different nodes. (Waits can also be caused by simultaneous updates within the same node, but these waits have nothing to do with replication.)

On the other hand, there are no replication-induced deadlocks with asynchronous replication. There is the possibility that an asynchronously replicated transaction will deadlock with another transaction within its own node, and this case is considered later.

Note that deadlocks require two waits, one for each of the opposing applications. Thus, *if waits (and therefore collisions) are rare, deadlocks will be extremely rare.*[35]
This is shown by example later.

This chapter analyzes expected wait, deadlock, and collision rates. It is based on work by Gray, et al.,[35] and extends their work in several important dimensions:

(1) Gray, et al., assumed that replication latency time for synchronous replication is zero. This analysis includes the effect of replication latency.

---

[35] Gray, J.; et al.; "The Dangers of Replication and a Solution," ACM SIGMOD 96, pp. 173-182; June, 1996.

(2) Gray, et al., assumed that all nodes contain a database copy. This analysis includes the case of *k* processing nodes, only *d* of which contain a database.

(3) This analysis is extended to cover those cases in which not all database modifications will cause a collision.

(4) This analysis shows how to analyze file/table hot spots.

The Model

The systems modeled follow the architecture shown in Figure 9-1. A system comprises a group of *processing nodes*. Some of these processing nodes also contain a copy of the database. A processing node containing a copy of the database is called a *database node*. Included is the case in which the database is partitioned, with each partition spread across several nodes. Each processing node (i.e., a node without a database) is associated with a specific database node containing the database (or with a set of database nodes containing the various database partitions).

The following parameters are used:

- k = total number of processing and database nodes.
- d = number of database nodes.
- r = application transaction rate arriving at each node (processing or database).
- a = for asynchronous replication, the number of modification actions in an average transaction that can cause a collision (inserts, updates, deletes).
  = for synchronous replication, the total number of modification actions in a transaction.
- a' = the total number of modification actions in an average transaction (whether or not they can cause a collision under asynchronous replication). For synchronous replication, $a = a'$.
- t = average time required to complete an action. Includes communication time for dual writes. Excludes

Breaking the Availability Barrier

communication time for synchronous replication using coordinated commits and for asynchronous replication.

L = replication latency time.

D = number of lockable items in the database (for instance, the number of records or rows) independent of the number of database copies. For instance, if the database contains 10,000,000 rows, and if there are two copies of that database in the system, then $D = 10,000,000$. The term *data item* in the following analysis is synonymous with *lockable item*.

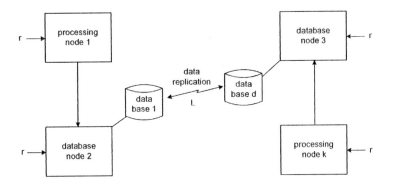

General Replication Model
Figure 9-1

In the following analysis, the terms "action" and "modification" are used interchangeably. That is, an *action* is a *modification* to a data item.

The following cases are considered:

(1) Synchronous Replication deadlocks caused by:

    a) mutual waits.

    b) lock latency.

(2) Collisions under Asynchronous Replication, where:

    a) Replicated transactions are sent serially to the remote databases one at a time following the commit at the source database.

    b) Replicated transactions are broadcast to all remote databases following the commit at the source database.

    c) Replicated modifications are sent serially to the remote databases as soon as they have been applied to the source database.

    d) Replicated modifications are broadcast to all remote databases as soon as they have been applied to the source database.

    e) Replicated modifications are broadcast to all remote databases as soon as they are received.

Note that each of these asynchronous techniques reduces the replication latency relative to the previous technique. The shorter the replication latency, the less chance there is for collisions.

These analyses are first done for the case of all nodes containing a database copy. That is, all nodes are database nodes ($k = d$). The results are then extended to the case for which only some nodes are database nodes ($k > d$). Further extensions cover the cases of transaction actions which do not cause collisions and of file or table hot spots.

Model Summary

The model results are summarized below, where

R_w = transaction wait rate (synchronous replication).
R_d = transaction deadlock rate (synchronous replication).
R_c = transaction collision rate (asynchronous replication).

Equation numbers are noted to help locate the derivation of each Equation in the following analyses.

The remainder of this chapter deals with the development of these relations. It is assumed throughout that waits, collisions, and deadlocks are rare so that second order effects can be ignored. It is also assumed that transactions occur randomly and are unrelated. That is, there is no batching of transactions, nor are there file or table hot spots.

Synchronous Replication

Mutual Waits

$$R_w = \frac{d(kra)^2 at}{2D} \qquad (9\text{-}26)$$

$$R_d = \frac{d(kra)^2 a^3 t}{4D^2} \qquad (9\text{-}27)$$

Note: For dual writes, action time t includes communication time.
For coordinated commits, action time t does not include communication time.

Lock Latency

$$R_d = \left(\frac{d-1}{d}\right)\frac{(kra)^2}{D} S \qquad (9\text{-}28)$$

Note: For dual writes, latency time, L, is the communication channel time. For coordinated commits, L is the replication channel latency.

Asynchronous Replication

$$R_c = \left(\frac{d-1}{d}\right)\frac{(kra)^2}{D}S \qquad (9\text{-}29)$$

where S is the stale time. Stale time is the time that a data item at one node is out-of-date because it has been updated at another node, and that update has not yet been replicated. In the evaluations of stale time, S, given below, the action time, t, excludes communication time.

Transactions Sent Serially After Commit

$$S = \frac{d(a't+L)}{2} \qquad (9\text{-}30))$$

Transactions Broadcast After Commit

$$S = a't + L \qquad (9\text{-}31)$$

Modifications Sent Serially After Application

$$S = \frac{d(t+L)}{2} \qquad (9\text{-}21)$$

Modifications Broadcast After Application

$$S = t + L \qquad (9\text{-}22)$$

Modifications Broadcast Upon Receipt

$$S = L \qquad (9\text{-}23)$$

Mutual Wait Deadlocks

We first analyze deadlock rates for co-located systems in which there are no delays associated with inter-node replication (that is, there is no stale time). Therefore, there are no deadlocks associated with lock latency. There are only mutual wait deadlocks.

When a transaction is synchronously replicated, each database modification is applied to all database copies within the scope of a common transaction. Only when all modifications across the network have been completed is the transaction committed and the modifications made permanent. The analysis for this case is a restatement of the analysis made by Gray, et al.[35]

We assume that modifications made by the transaction are executed serially. For instance, if a data item is to be modified, it is modified at the source database first and the modifications are then sent one at a time to each of the remote databases (serial writes). In some systems, all updates may be made simultaneously across the network (parallel writes). The following analysis can be easily modified to account for parallel writes, as is noted below with respect to Equation (9-2). All updates at all database copies are committed simultaneously once they are all ready.

We also assume that all nodes are database nodes ($k = d$), that database updates are made randomly across the database, and that there is no particular order in which a transaction applies its modifications (no ILP). Clearly, if there is an effective ILP, there will be no deadlocks. The resolution of deadlocks is described in the later section entitled *"Deadlock Resolution."*

What we do consider here is the probability that a pair of transactions on the same or different nodes in the application network will each have to wait on a data item locked by the other – the deadlock case of *mutual waits*.

We ignore the possibility that a node will lock a data item locally before it receives notification that a remote node has already locked that data item. That is, there is no lock latency. Deadlocks due to lock latency are considered later. (Note: Some distributed transaction managers may grant a lock on a data item only if they can acquire locks on all copies; otherwise, the requesting application will have to wait until locks on all copies can be acquired. This is the true case of zero lock latency.)

Transactions arrive at each node at a rate of r transactions/second. Since there are d nodes, the total system transaction rate is dr:

$$\text{system transaction rate} = dr \qquad (9\text{-}1)$$

Each transaction makes a modifications at d nodes and therefore has a lifetime of dat, where t is the average action time (including communication time in this case):

$$\text{transaction duration} = dat \qquad (9\text{-}2)$$

Note that for parallel writes, the transaction duration is only at. The following analysis can be modified for parallel writes by making this simple substitution. The result is that the relationships below for serial writes will all be divided by d in order to determine the relationships for parallel writes.

The total number of transactions in the system at any point in time is

$$\begin{aligned}
\text{concurrent transactions} &= \text{system transaction rate} \times \text{transaction duration} \\
&= (dr)(dat) \\
&= d^2 rat \qquad (9\text{-}3)
\end{aligned}$$

(If this relationship for concurrent transactions seems obscure, try putting in some values. If transactions are arriving at the system at a rate of 100 transactions per second, and if each transaction is 0.1 seconds long, then on the average, there must be ten transactions active in the system at any one time.)

Also on the average, each transaction will have $a/2$ data items locked. The total number of unique data items that are locked is

$$\begin{aligned}\text{total unique data items locked} &= \text{concurrent transactions} \times \text{actions}/2 \\ &= (d^2rat)(a/2) \\ &= d^2ra^2t/2 \end{aligned} \quad (9\text{-}4)$$

Since there are D lockable data items in the database, then

$$\begin{aligned}\text{probability that a particular data item is locked} &= \text{total unique data items locked/total data items} \\ &= (d^2ra^2t/2)/D \\ &= d^2ra^2t/2D \end{aligned} \quad (9\text{-}5)$$

A new action will wait if the data item that it wishes to modify is locked. The probability that this will happen is given by Equation (9-5). Thus,

$$\begin{aligned}\text{probability of an action wait} &= \text{probability that a particular data item is locked} \\ &= d^2ra^2t/2D \end{aligned} \quad (9\text{-}6)$$

A transaction will have to wait if any of its actions must wait. Assuming waits are rare, then

$$\begin{aligned}\text{probability of a transaction wait} &= \text{probability of an action wait} \times \text{actions} \\ &= (d^2ra^2t/2D)(a) \\ &= d^2ra^3t/2D \end{aligned} \quad (9\text{-}7)$$

Since transactions are arriving at a rate of dr, then

$$\begin{aligned}\text{transaction wait rate} &= \text{probability of a transaction wait} \times \text{transaction rate} \\ &= (d^2ra^3t/2D)(dr) \\ &= d^3r^2a^3t/2D \end{aligned} \quad (9\text{-}8)$$

A deadlock is caused when two transactions are waiting on each other. The probability that a particular transaction will deadlock with another specific transaction is the probability that the first transaction is waiting on an action locked by the second transaction <u>and</u> that the second transaction is waiting on another action locked by the first transaction. Equation (9-7) gives the probability that a transaction will wait on some other transaction. The probability that it will have to wait on some *particular* transaction is, using Equations (9-7) and (9-3),

> probability of a transaction wait on a specific transaction
> = probability of a transaction wait/number of transactions in the system
> = $(d^2ra^3t/2D)/(d^2rat)$
> = $a^2/2D$ (9-9)

Recalling our assumption that no intelligent locking protocol is used so that any action can potentially cause a deadlock, then

> probability of deadlock between two specific transactions
> = (probability of a transaction wait on a specific transaction)^2
> = $a^4/4D^2$ (9-10)

Using Equation (9-3) again, the probability that one of these transactions will deadlock on some other transaction is

> probability that a transaction will deadlock
> = probability of deadlock between two specific transactions x concurrent transactions
> = $(a^4/4D^2)(d^2rat)$
> = $d^2ra^5t/4D^2$ (9-11)

The deadlock rate is

> transaction deadlock rate
> = probability that a transaction will deadlock x transaction rate

$$= (d^2ra^5t/4D^2)(dr)$$
$$= d^3r^2a^5t/4D^2 \qquad (9\text{-}12)$$

The results for transaction wait rate (Equation (9-8)) and for transaction deadlock rate (Equation (9-12)) are the same as those obtained by Gray, et al. (their Equations (10) and (12)), albeit by a somewhat different route. In summary, the key results are the wait rate and the collision rate. Letting

R_w = transaction wait rate.
R_d = transaction deadlock rate.

then

$$R_w = \frac{d^3r^2a^3t}{2D} \qquad (9\text{-}8)$$

$$R_d = \frac{d^3r^2a^5t}{4D^2} \qquad (9\text{-}12)$$

Replication Conflicts

In the following cases, replication latency is introduced. This analysis is approached somewhat differently than it was for mutual waits. We assume that all nodes are database nodes ($k = d$) and that all actions within a transaction can potentially cause a conflict. For instance, there is no partitioning of the database.

Collisions Under Asynchronous Replication

When a transaction first modifies a data item, the value of that data item will be stale at a remote node until the data item at that node has received the lock. If one of these remote nodes attempts to modify that data item during its stale time, a collision results.

Consider a node A that has modified a data item during its stale time. Let

S = stale time

During this stale time, the number of data item modifications that will be made by applications running on node A is

    node modifications made during stale time
        = node application transaction rate x actions x stale time
        = raS   (9-13)

There are d-1 nodes remote to node A. Therefore, the total number of modifications made at remote nodes during the stale time is

    remote node modifications during stale time
        = number of remote nodes x node modifications made during stale time
        = (d-1) raS   (9-14)

The probability that a modification in node A will collide with a modification made at a remote node during the stale time is

    probability of a specific action collision
        = remote node modifications during stale time/number of data items
        = (d-1) raS/D   (9-15)

During the stale time, node A has made *raS* modifications. The probability that one of these will cause a collision is

    probability of a node collision during stale time
        = probability of a specific action collision x node modifications made during stale time
        = [(d-1)raS/D](raS)
        = (d-1)(raS)^2/D   (9-16)

Since there are d nodes, then the system-wide probability of a system collision is

probability of collision during stale time
= number of nodes x probability of a node collision during stale time
= d(d-1)(raS)²/D   (9-17)

This is the number of collisions that will be generated during the stale time. The collision rate is then

collision rate
= probability of collision during stale time/stale time
= [d(d-1)(raS)²/D]/S

or

$$R_c = \frac{d(d-1)r^2a^2}{D} S \qquad (9\text{-}18)$$

This is the result obtained by Gray, et al., (their Equation (18) for mobile applications), where their "disconnect time" is equivalent to stale time.

It now remains to calculate the stale time S for the various cases to be considered. We assume that stale time terminates as soon as the stale data item is locked because once it is locked, the local application cannot access it and cause a collision.

Transactions Sent Serially After Commit

If transactions are replicated serially following the source commit, then the stale time begins with the modification being made at the source node. On the average, $a/2$ more modifications must be made before the transaction is committed, requiring a time of $at/2$. The transaction must then be replicated to the other $d-1$ nodes one at a time. The last remote node to be updated will have to wait for the other $d-2$ remote nodes to be modified, requiring a time of $(d-2)(L+at)$. Thus, on the average, any remote node will wait a time of $(d-2)(L+at)/2$ for the other remote nodes to be modified. The remote node must then receive and apply the modifications to its database.

This will require an average time of $L+at/2$ to lock the stale data item in question. Thus,

$$S = \frac{at}{2} + \frac{(d-2)(L+at)}{2} + \left(L+\frac{at}{2}\right) = \frac{d(at+L)}{2} \quad (9\text{-}19)$$

Applying this result to Equation (9-18) for the case of zero latency ($L = 0$), we see that this is almost identical to the wait rate, R_w, for the case of synchronous replication (Equation (9-8)) except that one d term has been replaced by d-1 (an action can wait on another action in all d nodes, including its own node, under synchronous replication but will collide only with actions in the d-1 remote nodes under asynchronous replication). In fact, the two analyses proceeded along similar lines. This stresses that *waits under synchronous replication become collisions under asynchronous replication.*

Transactions Broadcast After Commit

Since there are a actions in a transaction, then when a data item is modified by the source transaction, there are an average of $a/2$ more data items to modify before the transaction commits This requires a time of $at/2$ until the commit. Once the transaction commits, it takes another L seconds for the remote nodes to get the new data item values followed by an average time of $at/2$ for the data item in question to be locked at the remote nodes. Thus,

$$S = \frac{at}{2} + \left(L+\frac{at}{2}\right) = at+L \quad (9\text{-}20)$$

Modifications Sent Serially After Application

The source modification will take a time t to be applied. The modification is then sent to d-1 nodes to be applied. Each remote modification will take a time L to be propagated to the remote node plus a time t to be applied. Following the argument presented in the section above entitled *"Transactions Sent Serially After Commit,"* the stale time is the sum of the source database modification time, t; one-half of the time for the previous remote nodes to be updated, $(d-2)(L+t)/2$; and the time for the data item to be locked at the remote node (this time is the replication time, L):

$$S = t + \frac{(d-2)(L+t)}{2} + L = \frac{d(t+L)}{2} \qquad (9\text{-}21)$$

Modifications Broadcast After Application

If modifications are sent to all nodes simultaneously after they are applied to the local node, the stale time is the time t to apply the source modification plus the time L to then send that modification to a remote node and to lock the data item:

$$S = t + L \qquad (9\text{-}22)$$

Modifications Broadcast Upon Receipt

If a modification is sent simultaneously to all nodes before it is applied locally, then the stale time is only the replication latency time until the data item is locked at the remote nodes:

$$S = L \qquad (9\text{-}23)$$

Deadlocks Under Synchronous Replication

In a multi-node system that uses synchronous replication to synchronize database copies, there are two ways in which a deadlock may occur when the latency of the replication channel is considered – mutual waits and lock latency.

Mutual Waits

If an intelligent locking protocol (ILP) is not used, then a deadlock may occur if two transactions must each wait on a data item locked by the other, as described earlier. The deadlock rate for this case has been previously analyzed and is given by Equation (9-12):

$$R_d = \frac{d^3 r^2 a^5 t}{4D^2} \qquad (9\text{-}12)$$

For dual writes, the action time t includes the communication time to send the action from one system to another. For coordinated commits, t is only the local action time.

Lock Latency

When lock latency is a factor, there is an interval from the time that a data item is locked in one node to the time that node's lock request appears at another node. This has been denoted as the stale time, S. During this time, a second node may also locally lock that same data item and attempt to acquire the lock on the data item at the first node. In this case, neither node can acquire the lock on the node remote to it. They will both wait on their respective remote node lock requests, and a deadlock results.

Under asynchronous replication, this is considered a collision. Therefore, when replication latency is a factor, we have a parallel to the rule which we noted previously: *A deadlock under synchronous replication becomes a collision under asynchronous replication when there is replication latency.* This is the lock latency deadlock that we have described earlier.

The lock latency deadlock rate, R_d, is therefore the same as the collision rate under asynchronous replication and is given by Equation (9-18):

$$R_d = \frac{d(d-1)r^2 a^2}{D} S \qquad (9\text{-}18)$$

In many synchronous replication engines, each update is forwarded immediately; and the stale time S will be the replication latency time L. For dual writes, S is simply the communication channel time. In any event, the following is clear:

Breaking the Availability Barrier

Rule 31: *Minimize lock latency to minimize synchronous replication deadlocks.*

Collisions, Waits, and Deadlocks

We had noted that waits and deadlocks in a distributed synchronously replicated system translate into collisions under asynchronous replication. This is illustrated in Figure 9-2, which shows an update on a source node being replicated to a target node.

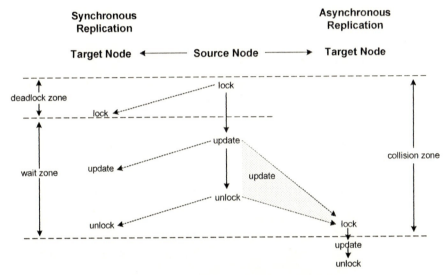

**Collisions, Waits, and Deadlocks
Figure 9-2**

Whether an asynchronous collision is equivalent to a synchronous wait or to a synchronous deadlock depends on when the data conflict occurs. Figure 9-2 shows an update being made at a node which we will call the *source node*. This update is being replicated to a node which we will call the *target node*. The path to the left represents the basics of synchronous replication (it would be a little different if dual writes or coordinated commits were to be shown more accurately, but it suffices to illustrate the points to be made). The path to the right shows the basics of asynchronous replication.

Let us take the synchronous replication case first:

a) <u>Deadlocks</u>: There is a delay between the time that a lock on a data item is acquired on the source node and the time that it is acquired on the target node. This delay is the lock latency described earlier. Should an application (or another replication engine) at the target node acquire a lock on that same data item during the lock latency period, a deadlock occurs because neither node can then acquire its lock on the other node. Note that such a deadlock is much more likely with coordinated commits than it is with dual writes because of the difference in lock latency times. For dual writes, lock latency is approximately the communication channel delay plus the time to acquire the lock on the target node. For coordinated commits, the lock latency is substantially the replication latency (which is communication time plus the replication engine processing time).

b) <u>Waits</u>: However, should the target node attempt to acquire a lock on the data item in question after the source node's lock has been propagated to the target node, then it will have to wait until the source node has released the lock on that data item. If updates are being replicated, this will occur when the data item is updated. However, if transactions are being replicated, then this will occur when the transaction is committed by the source node.

Let us now consider what happens under asynchronous replication.

c) <u>Collisions:</u> If updates or transactions are being asynchronously replicated, then the target node does not know about the data item lock until the update or transaction is replicated to it. The period from the time that a data item is locked at the source node to the time that it is locked at the target node is called the stale time. If the target node attempts to lock the data item in question during the stale time, it will succeed. Now both the source and target nodes may independently update the data item and replicate it to each other, thereby creating a collision. The asynchronous stale time encompasses both the synchronous lock latency and wait times. The length of the stale time depends upon whether updates or transactions are being replicated, as was analyzed earlier.

Thus, if during the lock latency time the target node attempts to update a data item locked by the source node, this will be a deadlock under synchronous replication but a collision under asynchronous replication. If the target node attempts to lock a data item locked at the source node after the lock latency time but before the time that the data item is unlocked at the target node under synchronous replication, then this will be a wait under synchronous replication but will still be a collision under asynchronous replication.

The fact that synchronous waits map to asynchronous collisions was our Rule 16 given in Chapter 3. We can now add a companion rule to Rule 16:

> **Rule 16:** (Gray's Law) - *Waits under synchronous replication become data collisions under asynchronous replication.*
>
> **Rule 32:** *Lock latency deadlocks under synchronous replication become collisions under asynchronous replication.*

Combined Effects

a) <u>Synchronous Replication</u>

Combining the mutual wait and lock latency deadlock mechanisms represented by Equations (9-12) and (9-18), the total deadlock wait for synchronous replication, R_d, is

$$R_d = \frac{d^3 r^2 a^5 t}{4D^2} + \frac{d(d-1)r^2 a^2}{D} S \quad \text{for sync. repl.} \quad (9\text{-}24)$$

Again, S is the lock latency time. For dual writes, it is the communication channel time plus lock acquisition time. For coordinated commits, it is the replication latency time (given by

Equations (9-19) through (9-23), depending upon the replication strategy).

Note that as the lock latency time, S, becomes very small, the second term of Equation (9-24) disappears; and the deadlock rate approaches that of mutual waits. However, even for very small values of S, the second term quickly becomes predominant because of the denominator D^2 in the first term (D is the number of lockable items in the database and is very large). Therefore, in the normal case, the deadlock rate is governed by lock latency rather than by mutual waits; and the first term in Equation (9-24) can be ignored. (The exception to this case occurs if a distributed lock manager which eliminates lock latency is used).

b) <u>Asynchronous Replication</u>

During asynchronous replication, there are no deadlocks caused by lock latency. However, there may be deadlocks caused by mutual waits with applications or other replication engines within the same node if an ILP is not used. The deadlock rate, R_d, can be determined from Equation (9-12) by setting $d = 1$ to obtain the deadlock rate within a single node and by then multiplying the result by d since there are d nodes within which deadlocks may occur. Thus, asynchronous replication deadlock rate is

$$R_d = d \frac{r^2 a^5 t}{4D^2} \qquad \text{for async. repl.} \qquad (9\text{-}25)$$

The collision rate for an asynchronously replicated system has been previously derived and is

$$R_c = \frac{d(d-1)r^2 a^2}{D} S \qquad (9\text{-}18)$$

where S is the asynchronous replication stale time given by Equations (9-19) through (9-23) for various replication engine configurations.

Breaking the Availability Barrier

For asynchronous replication, the deadlock rate will be orders of magnitude less than the collision rate because of the nature of the denominators in the above relationships, as described previously.

Although the expression for asynchronous collisions is the same as for synchronous lock latency deadlocks, the former is generally greater than the latter. Lock latency time is typically no longer than an update latency, whereas asynchronous stale time is typically at least equal to an update latency and may be as long as a transaction latency, depending upon the nature of the asynchronous replication engine.

Deadlock Resolution

It has now been shown that deadlocks are a major concern for synchronous replication, whereas collisions are a major concern for asynchronous replication. However, deadlocks are much easier to resolve than collisions (collision detection and avoidance are discussed in Chapter 3, "*Asynchronous Replication.*") In fact, there are several ways to either avoid deadlocks (even under asynchronous replication) or to resolve them automatically.

- Intelligent Locking Protocol: In a single node system, there are intelligent locking protocols (ILPs) that guarantee there will be no deadlocks. For example, records or rows can always be locked in a specific order. In this case, if a second transaction must wait on a data item locked by a first transaction, then the second transaction cannot lock a data item that will subsequently block the first transaction.

    Alternatively, it might be required that a transaction always acquires a record or a row that acts as a *mutex* (a mutually exclusive object) before it locks rows or records protected by that mutex. For instance, it might be required that a lock be acquired on an order header before any of its order detail records can be locked.

    In a synchronously replicated system, following an ILP in each of the independent nodes does not guarantee the avoidance of deadlocks due to lock latency. Locks may be acquired on the same data items in different database copies during the lock latency, thus creating a deadlock.

However, a modification to this procedure will prevent deadlocks in a distributed system. If one node is designated to be the master node, and if all locks are acquired first on the master node according to an ILP, then deadlock avoidance is guaranteed:

Rule 33: *Designating a master node for lock coordination can eliminate lock latency deadlocks when using synchronous replication.*

Of course, the use of an ILP will prevent deadlocks among applications on the same node if all applications are following the same ILP. Since this is the only source of deadlocks for an asynchronous system, then there will be no deadlocks under asynchronous replication if the same ILP is followed.

- Timeouts: If a deadlock does occur, then the application can decide after a time interval that a deadlock has occurred and can abort the transaction. This will release the other transaction so that it can complete. The aborted transaction can then be re-processed.

- Partitioned Sequence Numbers: A common database hotspot is a row in a table used to assign sequence numbers, such as for purchase orders. In a distributed synchronously replicated application, deadlocks on this row may be quite common due to lock latency. One way to avoid such deadlocks is to partition the sequence numbers among the various systems in the network. For instance, if there are n systems, system i is assigned sequence numbers $i + kn$, where k is incremented for each new sequence number. Thus, if there are three systems, System 0 will use sequence numbers 0, 3, 6, ..., System 1 will use sequence numbers 1, 4, 7, ..., and System 2 will use sequence numbers 2, 5, 8 ... Using this algorithm, the sequence number assignment table does not even need to be replicated. Each system can maintain its own sequence number table for its own use.

Breaking the Availability Barrier

Alternatively, the sequence numbers can be divided into non-overlapping ranges, with each node in the network assigned one of these ranges. Note that this numbering algorithm is applicable as well to the prevention of data collisions in asynchronously replicated systems.

- Synchronization Points: For coordinated commits, the source node can periodically ask the target node if it has acquired all locks so far. If the target node responds negatively, the source node can abort the transaction. Alternatively, the target node can independently notify the source node if it is unable to obtain a lock so that the source node can abort the transaction. Either of these techniques allows a potentially deadlocked transaction to be aborted part way through.

- Lock Table: Alternatively, a coordinated commit replication engine can use a lock table common to all nodes to coordinate locks. Prior to locking a record, the replication engine checks this common table to see if the record or row is already locked. It so, it waits until the lock frees so that it can acquire it. Once acquired, an indication to that effect is inserted into the lock table.

Note that the use of a lock table compromises the efficiency of coordinated commits. Under coordinated commits, an application proceeds at single-node speed until it has to wait for a ready-to-commit acknowledgement from the target node. If a common lock table is used, each database access is slowed down by the requirement to access the lock table. Furthermore, the advantage of non-invasive retrofitting of the application for replication may be lost.

Dr. Bill Highleyman, Paul J. Holenstein, and Dr. Bruce Holenstein

Processing Nodes and Database Nodes

So far, our analysis has assumed that all nodes are database nodes. However, in the general case, not all nodes are database nodes. There are k processing nodes and d database nodes (processing nodes with an attached copy of the database).

To reflect this more general case, the architecture of Figure 9-1 can be reduced to a system containing only database nodes, as shown in Figure 9-3. Generally, each processing node will be assigned to some specific database node to which it will route its database requests as it executes its transactions. A processing node's remote database access services to its associated database node will be provided by an RPC or some equivalent mechanism. It is assumed that the time required for such remote access can either be ignored or is included in the average action time.

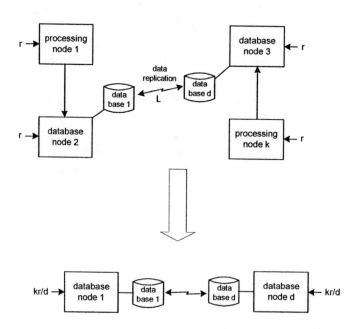

**General Model Reduced to Database Node Only Model
Figure 9-3**

A database node and its associated processing nodes can then be considered as a single database node with an incoming transaction rate of kr/d – i.e., $1/d$ of the total system transaction rate kr.

Thus, the enhancement of the previous analyses to cover cases in which not all nodes are database nodes is to simply replace the term r in each of the relationships with the expression kr/d. This results in the following relationships:

(1) <u>Synchronous Replication</u>

   a) Mutual Waits:

   From Equation (9-8),
   $$R_w = \frac{d^3 \left(\frac{kr}{d}\right)^2 a^3 t}{2D} = \frac{d(kra)^2 at}{2D} \qquad (9\text{-}26)$$

   From Equation (9-12),
   $$R_d = \frac{d^3 \left(\frac{kr}{d}\right)^2 a^5 t}{4D^2} = \frac{d(kra)^2 a^3 t}{4D^2} \qquad (9\text{-}27)$$

   b) Lock Latency:

   From Equation (9-18),
   $$R_d = \frac{d(d-1)\left(\frac{kr}{d}\right)^2 a^2}{D} S = \left(\frac{d-1}{d}\right)\frac{(kra)^2}{D} S \qquad (9\text{-}28)$$

(2) <u>Asynchronous Replication</u>

   $$R_c = \frac{d(d-1)\left(\frac{kr}{d}\right)^2 a^2}{D} S = \left(\frac{d-1}{d}\right)\frac{(kra)^2}{D} S \qquad (9\text{-}29)$$

where S is given by Equations (9-19) through (9-23).

Not All Actions Collide

It is quite possible that not all actions in a transaction will collide. Some may be made to files or tables for which collision is not a possibility or is not a concern. Others, such as inserts, are generally not candidates for collisions. These actions will contribute to stale time because they extend the length of a transaction, but they do not contribute to the probability of a data collision.

Reviewing the derivation of Equation (9-28), which evolved from Equation (9-18), one can verify that the term a^2 contributes to the probability of collisions and therefore represents the actions that can cause a collision. However, the term S has only to do with the amount of time during which a collision may occur.

Therefore, to evaluate the collision rate, the term a^2 in Equation (9-18) is based on the number of transaction actions that can cause a collision, whereas any term a that is included in the evaluation of the stale time, S, represents all actions in a transaction, whether they cause a collision or not. Let

a = number of transaction actions that may cause a collision.
a' = total number of actions in a transaction.

Equation (9-28) above remains unchanged, but the stale time for the asynchronous transaction replication cases, Equations (9-18) and (9-19), become

$$S = \frac{d(a't+L)}{2} \quad \text{(trans. sent serially after commit)} \quad (9\text{-}30)$$

$$S = a't+L \quad \text{(trans. broadcast after commit)} \quad (9\text{-}31)$$

File/Table Hot Spots

In most applications, collisions can occur across many files or tables. Each of these files or tables may be of a different size and may have different action rates applied against them. If a file or table is small compared to the other files or tables in the application, and/or if the action rate against that file or table is higher than average compared to the other files or tables, then this file or table may become a *hot spot*. It may account for the bulk of collisions.

The above analysis has suggested that the total database size be used in the calculations along with the total action rate. However, a more accurate approach is to apply the above equations to each file or table individually to determine the collision rate across each and then to sum these collision rates to get the total application collision rate. By doing so, the hot spots will be made apparent.

Examples

A feel for the relative frequency of deadlocks, waits, and collisions can be obtained by example. Let us consider the following system:

4	database nodes ($d = k = 4$)
10	transactions per second per node ($r = 10$)
4	actions per transaction ($a = 4$)
20	msec. action time ($t = .02$)
100	msec. replication latency time ($L = .4$)
20	msec. communication channel time
10,000,000	lockable data items (rows) in the database ($D = 10,000,000$).

Dr. Bill Highleyman, Paul J. Holenstein, and Dr. Bruce Holenstein

Synchronous Replication

Dual Writes

If dual writes are used, action time must include communication channel time; so effective action time, t, is 40 msec. Lock latency time is also communication time, so S is 20 msec. Thus, wait and deadlock rates are

$$R_w = \frac{d^3 r^2 a^3 t}{2D} = 2.95 \text{ waits per hour}$$

$$R_d = \frac{d^3 r^2 a^5 t}{4D^2} + \left(\frac{(d-1)}{d}\right)\frac{(kra)^2}{D} S = .138 \text{ deadlocks per hour}$$

Note that deadlocks are substantially caused by lock latency. Mutual wait deadlocks are measured in centuries for this case (see the deadlock calculation for asynchronous replication below).

Coordinated Commits

For coordinated commits, let us assume that lock latency time is replication channel latency, so that $S = 100$ msec. Then the deadlock rate is

$$R_d = \left(\frac{(d-1)}{d}\right)\frac{(kra)^2}{D} S = .69 \text{ deadlocks per hour}$$

Note that this is five times the dual write deadlock rate because the lock latency time is five times as long (20 msec. communication channel time for dual writes versus 100 msec. stale time for coordinated commits).

Asynchronous Replication

For asynchronous replication, action time excludes communication time; and $t = 20$ msec. Let us assume that updates are broadcast to the target nodes after they are applied at the source node. Therefore, stale time is given by Equation (9-22):

$$S = t + L = 120 \text{ msec.}$$

Then the expected collision rate is

$$R_c = \left(\frac{(d-1)}{d}\right)\frac{(kra)^2}{D}S = .83 \text{ collisions per hour}$$

The expected deadlock rate due to same-node mutual wait deadlocks is

$$R_d = d\frac{r^2a^5t}{4D^2} = .0646 \text{ deadlocks/century}$$

Mutual wait deadlocks can be ignored for this case.

Chapter 10 - Referential Integrity

In Chapter 3, *"Asynchronous Replication,"* we discussed the importance of *natural flow* and the problems that might occur if database changes were not applied to the target database in the same order as they were applied to the source database. In this chapter, we look further at these ramifications and how different replication engine architectures can be configured to provide natural flow data replication.

Background

System Replication

As enterprises become more and more dependent upon their computing infrastructure, the continuous availability of these systems assumes an ever increasing importance. A powerful way in which to significantly increase a system's availability is to replicate it. That is, an independent backup system is provided, one that can continue to provide data processing services in the event of a failure of the primary system.

There are many ways to replicate a system, as shown in Figure 10-1:

a) *Cold Standby:* The backup system is not involved in the primary system's application unless the primary system fails. Should this happen, the failed applications are loaded into the backup system so that processing can resume under the control of the backup system. Cold standbys are usually used with tape backup. Periodic snapshots (and perhaps more frequent updates) are made of the primary system's database and held in safe storage. Should the primary system fail, the backup tapes are retrieved from storage and loaded onto the backup system so that the backup system can resume processing. In some cases, the backup tapes are preloaded onto the backup system to shorten the recovery time from a primary failure.

b) *Warm/Hot Standby*: As is the case with a cold standby, a backup system is provided, one which is not normally involved in the primary system's application. However, all of the primary system's applications are preloaded in the backup system so that it is prepared to quickly take over in the event of a primary failure. In some cases, the applications may have the files open only for read access. In these cases, following a primary failure, the applications must reopen their files for full access before they can take over the processing functions of the primary system. This is called a warm standby. A hot standby has all of its files open for full access and is ready to take over immediately following a primary system failure. In either case, the standby system may be used for read-only activities such as query or reporting, but it is not updating the database.

c) *Active/Active*: In this architecture, both systems are actively running the same application and are sharing the processing load. Should one node fail, application services continue to be provided by the surviving system without interruption except that total capacity has been cut in half.

d) *Multi-Node:* The active/active architecture can be extended to allow the application to run simultaneously on many processing nodes. If there are k nodes, then the failure of any one node will result in the loss of only $1/k$ of the total system capacity. For instance, the loss of one node of a four-node system will result in a 25% reduction in processing capacity.

Breaking the Availability Barrier

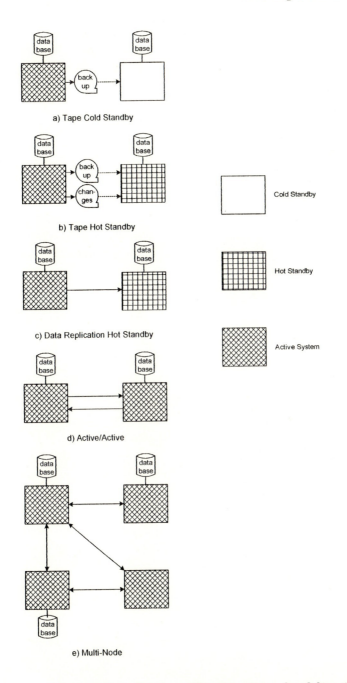

**Illustrative System Replication Architectures
Figure 10-1**

The *availability* of a system is the proportion of time that it will be operational. Typical systems today have availabilities of 99% to 99.99%. It is common to measure availability in terms of "9s." A system with 99% availability has an availability of two 9s. A system with 99.99% availability has an availability of four 9s. In a 24-hour per day, 7-day per week operation, a system with two 9s availability will be non-operational more than eighty hours per year on the average. A system with four 9s availability will be down on the average less than one hour per year.

It was shown in Chapter 1, *"The 9s Game,"* that replicating a system (that is, adding a backup system) will double the 9s of the system. If k systems are provided, then the resulting application network will have an availability that is k times the number of 9s as a single system. For instance, if a single system has an availability of three 9s, providing a replicate system will increase its availability to six 9s.

There are many reasons to replicate a system, including:

- to significantly increase the system's availability for its community of users.

- to provide tolerance to natural or man-made disasters by separating the nodes by large distances so that such a disaster will affect only one node.

- to allow maintenance or updating of the system one node at a time so that the application does not have to be taken down.

- to provide locality of operations and data to user groups in the field.

Data Replication

Providing replicated processing capacity is only part of the requirement to replicate a system. In order for an application to function properly, it must have access to the database that represents the current state of the system (such as part levels for an inventory system, account balances for a banking system, and so forth). Therefore, the applications running on each node in a replicated system must have access to a current copy of the database, as shown in Figure 10-1. Not only must the processing capacity be replicated, but the database must be replicated as well.

There are many ways to replicate a database, and each has its own characteristics with regard to recovery time following a failure and the amount of data that may be lost as a consequence of the failure. The means of data replication is chosen in part to conform to a company's tolerance for down time and data loss as determined by company objectives. These objectives are known as the Recovery Time Objective (RTO) and the Recovery Point Objective (RPO) as discussed in detail in Chapter 6, *"RPO and RTO."*

Early Systems

In early replicated systems, a cold standby architecture was used; and data replication was via magnetic tape. Periodic copies of the primary database were written to tape, and that tape was transported to the backup system and used to update that system's database. Backup might typically have been undertaken daily or weekly.

A significant problem with this procedure was that a great deal of data was lost whenever the primary failed. The backup system had available to it a database that was perhaps hours or even days old.

This problem was alleviated by writing to magnetic tape the changes that were made to the database as the changes occurred. These changes could then be applied to the backup database by transporting the change tape to the backup system and by applying the changes to the backup database prior to activating the backup application.

Although little data now was lost, the recovery time for the application could be measured in hours or even days while the change tapes were transported to the backup site and loaded. Furthermore, if the change tapes were destroyed in the disaster, or if they proved to be unreadable, then all those changes were lost.

Asynchronous Replication

To solve the problems inherent with tape backups, real-time data replication engines were developed. These engines replace the magnetic tape with a communication network. As changes are made at the primary database, they are communicated in near-real-time to the backup system, where they are immediately applied to the backup system's database. The backup system is typically run as a warm or a hot standby.

Since database modifications are being applied to the backup independently of the source database modifications (and at some time later than the source updates), this is called *asynchronous* data replication. Asynchronous data replication is discussed in detail in Chapter 3, "*Asynchronous Replication.*" Although this is very much faster than data backup using magnetic tape, there is some delay from the time that a change is made to the primary database and the time that it is applied to the backup database. This delay is called *replication latency.* Replication latency can be considered the time that it takes a change to propagate through the replication pipeline and to be applied to the target database. Changes in the replication pipeline at the time of a primary system failure will most likely be lost. However, this generally represents at most a few seconds of data; and recovery at a hot standby can be very fast.

The fact that the backup system is now in near synchronism with the primary system allows active/active architectures to be implemented. That is, applications may be active in all nodes of the replicated system, with each application having available to it a nearly up-to-date copy of the database.

The terms *primary* and *backup* databases do not apply to active/active systems since, in effect, every database copy in the network is backing up every other database copy. Rather, databases are referred to as *source* and *target* databases. A change is applied to a source database and is replicated to one or more target databases, which can themselves be source databases updating other target databases.

Synchronous Replication

If exact database copies are required or if no data loss is tolerable, there are techniques available that will guarantee that all database copies are identical and that no data will be lost. These techniques are generally known as *synchronous replication* since all changes are synchronized across the network. Synchronous replication is discussed in detail in Chapter 4, *"Synchronous Replication."*

Synchronous replication techniques may be dichotomized into weak synchronous replication and strong synchronous replication. Strong synchronous replication guarantees that the modifications are in fact applied to the target database when they are applied to the source database.

Weak synchronous replication guarantees only that the changes included in a transaction have been received by the target site, but it does not guarantee that they have been applied to the target database. At this point, the source system is free to commit its transaction. However, there is no guarantee that the changes that are safe-stored by the target system can in fact be subsequently applied to the target database. If pending transactions at the target system are aborted after the system has committed them, then the databases are no longer in synchronism.

In Chapter 4, *"Synchronous Replication,"* we discussed methods for coordinating commits across a network by using Ready to Commit (RTC) or Ready to Synchronize (RTS) tokens returned from the target node to the source node in response to a source node query.[36] These tokens can also be used to implement weak synchronization by using them to respond to a source system query to inform the source node that all of the modifications required by a transaction have been received and/or have been safely stored but have not necessarily been

---

[36] Holenstein, B. D., et al.; *"Collision Avoidance in Data Replication Systems,"* United States Patent Application No. 2002/0133507; Sept. 19, 2002.

applied to the target database. Upon receipt by the source node of this response, the source node's transaction can be allowed to complete.

Guaranteeing that database updates will be applied to the target database is also a problem with asynchronous data replication. In effect, weak synchronous data replication is a halfway step between strong synchronous data replication and asynchronous data replication.

Both strong and weak synchronous replication have performance issues which asynchronous replication does not have. Because each update must be coordinated among all copies of the database across the network, the application is slowed. The increased time that an application must take to provide this coordination is known as *application latency* and adds to the transaction response time.

Asynchronous replication does not affect transaction response time since the remote database updates are made without any application involvement. However, both asynchronous replication and weak synchronous replication may be subject to data collisions - described later - which strong synchronous replication avoids. Only the effects of asynchronous replication and weak synchronous replication are considered herein since there are generally no database consistency problems introduced by strong synchronous replication.

Physical Replication

There is another form of data replication which is mentioned here for completeness, and that is physical replication. Some systems will replicate data at the physical level. That is, whenever a disk block has changed or a time interval has expired on a changed block, that block is queued for transmission to the target system, where it is applied over the existing block at the target system. Physical replication has several limitations. They include:

- If there is data corruption due to some fault at the source node, then the corruption will be replicated to the target node.

- Physical replication ignores event and transaction ordering since blocks may not be sent in event order but according to some other algorithm.

- Physical replication ignores transaction boundaries, so inconsistent views of the database may be quite common.

- There is no guarantee that indices will match the data at any given point in time.

- Physical replication does not support heterogeneous replication - both the source and target databases and systems must be the same.

- A great deal of data that has not changed is sent over the communication line as part of each physical block.

On the other hand, physical replication can be much faster. It can play a role in unidirectional replication. However, since it provides no semblance of data consistency, it is hardly suitable for active/active applications and is considered no further herein.

Asynchronous Replication Issues

There are several issues associated with asynchronous data replication. First, as noted above, changes in the replication pipeline may be lost in the event of a source system or network failure. This data loss can be minimized by minimizing replication latency, and lost data is generally recoverable once the failure has been corrected (assuming that the source system has been recovered).

Second, the ping-ponging of changes in active/active applications must be avoided. Ping-ponging is the replication of a change received by a target system back to the source system. There are techniques available for ping-pong avoidance.[37] (Certain synchronous replication techniques are also subject to ping-ponging.)

---

[37] Strickler, G., et al.; *"Bi-directional Database Replication Scheme for Controlling Ping-Ponging;"* United States Patent 6,122,630; Sept. 19, 2000

Third, asynchronous data replication is subject to data collisions. Data collisions are caused by replication latency. Because there is a delay from the time that a change is made to the source system and the time that it is applied to the target system, there is a time interval during which the value of that data item at the target system is *stale*. That is, the value at the target system is incorrect. Should an application at the target system change a stale data item, then it will replicate the resulting value to all other systems in the network at the same time that the original source item is being replicated. Each of these replicated values will be different, and each probably will be wrong. This is known as a *data collision*, and the database is now in an inconsistent state. Data collisions must either be detected and corrected, or they must be avoided (see Chapter 3, "*Asynchronous Replication*," for a discussion of data collision detection and correction).

Data collisions can be avoided in one of several ways. For example, the database can be partitioned such that each node *owns* a partition; and only the owner of a partition can change that partition. If this is done, data collisions will not happen. If, however, all nodes must be able to change all data items, then synchronous replication can be used to guarantee that all copies of a data item are changed before any copy can be further changed.

If data collisions cannot be avoided, then they can cause the database to be placed into an inconsistent state. A fourth problem with asynchronous data replication is that there are other situations or configurations that can cause the database to become inconsistent. These situations are said to violate the *referential integrity* of the database. Referential integrity and its violation situations are described in more detail later.

Transactions

Fundamental to database consistency is the concept of a *transaction*. A transaction is a group of related changes that are managed in such a way as to maintain a database in a consistent state.

That is, a view of the database at any time will always give a consistent result.

A simple view of the problem solved by the transaction concept can be obtained by considering a banking application that is managing a person's checking account and savings account. If $1,000 is to be transferred from his savings account to his checking account, then that amount must first be subtracted from the savings account and then added to his checking account (or vice versa). If the balances for these accounts are viewed in the middle of this process, then the savings account balance will be reduced by $1,000, but this money will not have shown up yet in the checking account. Even worse, should the system fail at this point, then $1,000 will be lost.

By using the transaction model, this problem is avoided. A transaction manager assures that either all changes within the scope of a transaction are made (the transaction is *committed*) or that none are made (the transaction is *aborted*, thus returning the database to its original state). Additionally, the transaction manager usually assures that intermediate states are not readable.

Programmatically, the definition of a transaction's scope is accomplished by framing the changes comprising a transaction by a Begin Transaction command and an End Transaction command:

>Begin Transaction
>Change 1
>Change 2
>.
>.
>.
>Change n
>End Transaction

The transaction model has certain properties that are extremely important to database consistency. They are known as the ACID properties[38]:

Atomic – The transaction is an atomic entity. Either all changes are completely executed, or none are executed.

Consistent – Any view of the database at any time is always consistent. If a transaction is in process, one may see the pre-transaction data or be required to wait until the transaction completes, depending upon the system. But one will never see an intra-transaction inconsistent state. There is one common exception to consistency. Some database managers allow so-called "dirty reads" – the reading of locked data that may be in the process of being modified. Such reads may yield inconsistent results.

Isolated – The effects of a transaction are unaffected by other transactions that are being simultaneously executed.

Durable – The effects of a transaction survive system faults. There are many levels of durability. At the very least, the changed data must be written to a persistent storage device such as disk. That data will survive a disk failure if the disk is replicated (mirrored disk or RAID – Random Arrays of Independent Disks). The data will survive a disk failure and a processor failure if the system is replicated.

Transactions are effective across replicated systems with multiple copies of the database since the changes to each database copy may be included in the scope of the transaction. Thus, either all copies will be changed, or none will. Furthermore, the new data values at each copy will not be viewable until it is guaranteed that all copies have indeed been changed. This is synchronous replication, as discussed previously.

---

[38] Gray, J. et al.; <u>Transaction Processing: Concepts and Techniques</u>, Morgan Kaufman; 1993.

However, the transaction model loses some of its protective capabilities if asynchronous replication is used. This is because the replication process spawns independent transactions at each database copy. There is no coordination between these transactions. They each guarantee the ACID properties within their own database, but there is no such guarantee across the network.

Specifically, the database may have periods of inconsistency due to data collisions described previously or due to referential integrity violations, as described later.

Simple Data Replication Model

The basic components of an asynchronous data replication engine are shown in Figure 10-2. There are three components in this simple model.

- an *Extractor* that is responsible for obtaining changes to the source database.

- an *Applier* that is responsible for applying changes to the target database (the Applier can, in fact, be part of the same component that contains the Extractor).

- a *communication channel* that is used by the Extractor to send database changes to the Applier and to receive acknowledgement, control, and status information from the Applier.

As shown in Figure 10-2, the Extractor may obtain source database changes in any one of several ways. They include:

a) The application program generating the changes may queue change information for the Extractor to read. Though this queue may be memory-resident, it is usually written to a persistent store such as a disk-resident Change Log so that the changes are preserved in the event of a source system failure. (Note: The term "disk" will be used solely for convenience hereafter and is to be interpreted as "persistent storage" throughout this chapter.)

b) The transaction manager may create an Audit Trail on disk of all changes made to the database for the Extractor to read. The Audit Trail may comprise several disk volumes, such as a Master Audit Trail and one or more Auxiliary Audit Trails.

c) A change to the database may activate a database trigger, stored procedure, or publish/subscribe facility that will queue the change to the Extractor. Although this queue may be memory-resident, it is usually written to a disk-resident Database of Change (DOC) file for durability purposes in the event of a source system or network failure.

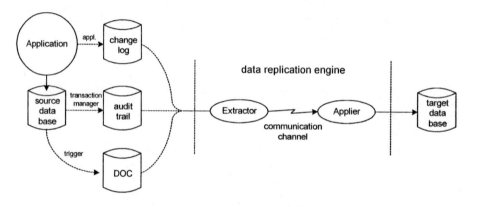

Simple Data Replication Engine
Figure 10-2

This simple data replication model is easily extended to bi-directional and multi-node data replication by using multiple data replication engines, as shown in Figure 10-3. It may also be extended to provide synchronous replication by adding a facility to coordinate the commits by the various replication engines across the network. Note that in a multi-node network, there need not be a connection between every node so long as each node has a path to all others. Also, there need not be a database resident at every node. Some nodes may have a partial database or no database at all; instead, they access data from other nodes. Data replication engines may only be needed at those nodes that have a partial or full database copy.

Breaking the Availability Barrier

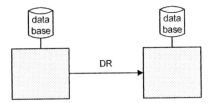

a) Unidirectional Replication

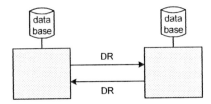

b) Bi-directional Replication

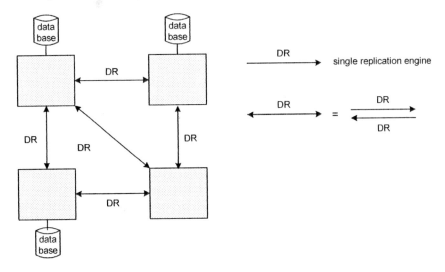

c) Multi-Node Replication

Data Replication Architectures
Figure 10-3

Natural Flow

In the simple data replication model shown in Figure 10-2, the replicated entity may either be a single data item change or a transaction. If changes are being replicated, then as each change is received by the Extractor, it is sent to the Applier and applied to the target database. Of course, these changes may be buffered or blocked to improve efficiency. If transactions are being replicated, then in a similar manner each begin, change, and end are given to the Extractor as they occur and are sent to the Applier for application to the target database.

The form of changes sent down the replication pipeline can vary from one data replication engine to another. Some will send after images of a row or record. Others will send just field or column changes within a row or record. Still others will send the operations or functions to be applied to a specific set of rows, fields, or columns or will send only the relative difference represented by the changes. Before images are often sent as well. Before images are useful to determine if a data collision has occurred by comparing the source before image to the current target image. Before images can also be used to back out modifications made by a transaction that is subsequently aborted.

In the simple model of Figure 10-2, the database change events at the source database are applied to the target database in precisely the same order. Thus, the target database represents an exact copy of the source database as it transitions through its various states, albeit delayed by the replication latency. Although the replication channel may be single-threaded, transactions may be interleaved as they were at the source system.

This simple model preserves the *natural flow* of changes or transactions occurring at the source database as they are being applied to the target database.[39] The requirement to adhere to strict event

---

[39] Knapp, H. W., "The Natural Flow of Transactions," ITI White Paper; 1996.

Breaking the Availability Barrier

sequence can be relaxed if events are not related. It can also be relaxed if database consistency is not checked until transaction commit time rather than upon each update. In this case, changes within a transaction can be applied out of order so long as database consistency is satisfied at commit time. These characteristics are taken advantage of in some of the architectures which follow.

If the natural flow of transactions is not preserved at the target database, then the database can become corrupted. This may occur for a variety of reasons. For instance, one transaction may set the value of a data item to 10. A subsequent transaction may set the value of that same data item to 20. If these transactions are applied in opposite order at the target database, then the data item will be left with a value of 10, which is wrong. The database is now corrupted.

A similar problem can happen even if the natural flow of transactions is preserved, but the change order within a transaction is not preserved. Corruption will occur if multiple operations within the transaction are not commutative – that is, the result of these operations depends upon the sequence in which they are executed. The simplest case is when the replication engine is sending changed images of rows, records, fields, or columns. In this case, if multiple changes have been made to a field or column or even to a row or record, an older image may overwrite a newer image if they are not applied in the correct order and will leave the database in an incorrect state.

A more subtle case occurs when only change operations are sent or are computed before the changes are applied. These operations are often commutative and can be applied in any order. But consider a data item A, which is to be modified by adding data item B to it and then multiplying the result by data item C:

$$(A+B)C \rightarrow A$$

If natural flow is not preserved, and if the multiplication is made before the addition, the result is

$$AC+B \rightarrow A$$

This is the wrong result. As will be demonstrated later, more complex data replication models may not inherently preserve natural flow. Such systems can cause database corruption unless special measures are taken to restore the natural flow prior to updating the target database.[40] *The purpose of our discussion here is to describe various measures to maintain natural flow.* The net result is concisely summarized in Rule 12 given in Chapter 3, *"Asynchronous Replication:"*

> **Rule 12:** *Database changes generally must be applied to the target database in natural flow order to prevent database corruption.*

Referential Integrity

Referential integrity is another property of a database that must be preserved in order for the database to be correct. A database often comprises a complex network of references between the various records or rows stored in its files or tables. It is generally of paramount importance that when a reference is made to a reference or row (or a field thereof), the referenced record or row actually exists and that the referenced entities are logically consistent (e.g., an accumulator value in one record or row equals the sum of the individual values in its referenced or referencing rows or records). Otherwise, the application may not be able to function correctly or may not be able to function at all. These references are sometimes known as *foreign keys*. All foreign keys must be resolved in order to satisfy referential integrity.

Figure 10-4 gives an example of the requirement for referential integrity. It shows an order processing database. An order is represented in the database by an Order Header row and one or more Order Detail rows. The Order Header row contains the order number and customer id for the order.

---

[40] There are some system implementations which do not care about database consistency unless there is a failover. In this case, the backup system will fix the database inconsistencies before taking over the processing load. These implementations do not support active/active applications.

Breaking the Availability Barrier

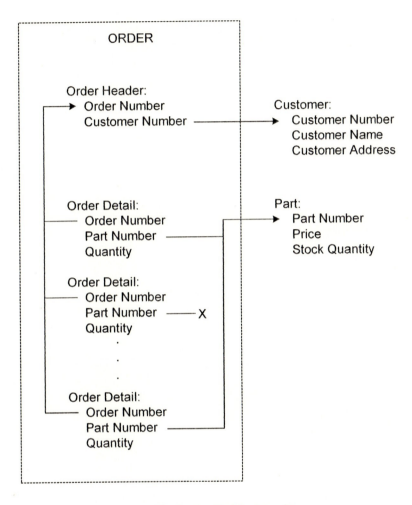

**Referential Integrity
Figure 10-4**

Each Order Detail row contains the order number and the part number and quantity required by the order for a particular part. Each part is represented in the database by a Part row that contains the part number, its price, and the quantity in stock.

Finally, there is a row for each customer; this row contains the customer number, contact information, and so forth.

In order to create and process an order, each Order Detail row must have access to its Order Header, which identifies the customer, and must also have access to its Part row to obtain pricing and availability information. In addition, the Order Header must have access to its Customer row in order to confirm the order and to send an invoice.

If the target row of any of these references, or foreign keys, does not exist, then the order cannot be processed. For instance, Figure 10-4 shows that the Part row for the second Order Detail row cannot be found. Therefore, the price and availability for this part cannot be determined; and either the order cannot be processed or can only be partially processed.

The existence of a target for every reference satisfies one of the conditions for *referential integrity*. If a reference is missing, the database has suffered a referential integrity violation. This is a foreign key that can not be resolved.

The record or row that is making the reference is known as the *child*. The record or row being referenced is the *parent*. A requirement for referential integrity, therefore, is that a parent must exist before its child is created; and a parent cannot be deleted until all of its children have been deleted.

There are also other application-specific constraints that may be violated if natural flow is not preserved. Ensuring that account balances do not become negative is one example. For instance, if one transaction adds an amount to an account, and if a second transaction debits that account, applying these transactions in the wrong order may result in a negative balance which may cause the debit transaction to be rejected at the target system even though it was accepted at the source system. This will not happen if natural flow is observed at the target system, nor will it happen if referential integrity is turned off at the target system.

Compressed audit trails may also cause referential integrity problems by interrupting natural flow. For instance, a change to a text field may just be recorded in the audit trail as the changed characters and a positional offset. In order to update the target row, the entire field or row must often be reconstructed. One way to do this is to apply the changed characters at the appropriate point to the target record or row. This works if the source and target databases have the same or similar structure but can be a difficult mapping task if the database structures are different.

Alternatively, the current source row can be read and replicated in addition to, or instead of, the compressed data and then applied to the target database. However, the contents of the source row may have changed in the meantime. Although the target result is correct after some point in time, natural flow may be violated if the source row contents corresponding to the change are not replicated.

The above examples show the importance of maintaining inter-transaction dependencies. It may also be important to maintain intra-transaction dependencies if referential integrity is checked upon each data item modification. Otherwise, the reversal of a debit and a credit or the reversal of a parent/child creation may cause transaction rejection due to referential integrity checks.

In addition, for our purposes, referential integrity requires that the ACID properties of transactions be maintained.

Thus, for our purposes herein, *referential integrity is taken to include the requirements that all foreign keys are resolved, that intra- and inter-transaction dependencies are preserved, and that there is no violation of the ACID properties of transactions.* In certain cases, some of these restrictions can be relaxed if they do not affect database consistency.

In the simple data replication model of Figure 10-2, the natural flow of changes or transactions is preserved. Therefore, the same level of referential integrity enforced at the source database will also be provided at the target database. If it is desired to have a higher level of referential integrity at the target database, then business rules can be incorporated into the Extractor, the Applier, or both to reorder transactions or events to provide this. This is especially important if the target system supports a stronger level of referential integrity than the source system, providing that that level of referential integrity is turned on at the target system.

Referential integrity may be enforced at the database level, depending upon the database being used. In fact, it may be enforced as each change is made; or it may be enforced only at transaction commit time. In the latter case, a change may violate referential integrity so long as the violation has been corrected at commit time. The enforcement of atomic changes (all or none) is usually an intra-transaction function, whereas the enforcement of other referential integrity criteria may be either an intra- or an inter-transaction function.

Thus, full referential integrity at the target database can only be guaranteed if natural flow is followed. Adding to Rule 12 which was repeated above, we have

> **Rule 34:** *Database changes must be applied to the target database in natural flow order to maintain referential integrity.*

Current Data Replication Architectures

There are several data replication architectures being used as of this writing. Many of these are shown in Figure 10-5. To simplify the following descriptions, the Audit Trail, DOC, or Change Log will be referred to generically as the Change Queue.

Single-Threaded Replication Engine

Figure 10-5a shows a basic single-threaded replication engine that was described earlier. Since all changes are sent to and applied to the target database in the same order as they were applied to the source database, then full natural flow order is maintained, at least so far as it was maintained in the Change Queue.

One complication with this model is that since all events are replicated, even those changes associated with transactions that are subsequently aborted are replicated and applied. The replication engine must have a means to remove these aborted transactions from the target database. This is typically done by the replication engine aborting the transaction at the target database, by applying the before images or undo events to the target database, or by the target system's transaction monitor performing similar tasks.

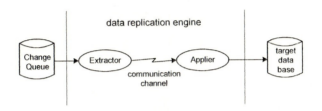

a) Single-Threaded Replication Engine

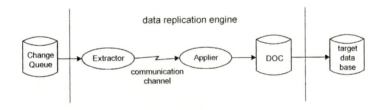

b) Single-Threaded Replication Engine with DOC

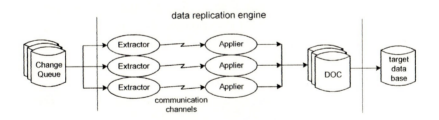

c) Multi-Threaded Replication Engine

**Current Data Replication Models
Figure 10-5**

Dr. Bill Highleyman, Paul J. Holenstein, and Dr. Bruce Holenstein

Single-Threaded Replication Engine With DOC

The abort complication with single-threaded replication engines is removed if a Database of Change (DOC) is used to buffer changes at the source or target system before they are applied to the target database, as shown in Figure 10-5b. The replication engine can now select from the DOC only those transactions that completed and can apply them to the target database. Aborted transactions are ignored. However, replication latency – the time that it takes for an update to the source database to propagate to the target database - has been lengthened by the extra storage step of transactions having to pass through the DOC intermediate storage.

Note that with a DOC, total natural flow is not followed since the modifications included within a transaction will not be applied to the target database until the transaction has committed. However, all change events within each transaction will be applied in natural flow order; and transactions themselves will be applied in the correct order with respect to each other. It is just that transactions are not interleaved as they are at the source system.

Also, to preserve transaction natural flow, the replication engine will generally apply just one transaction at a time from the DOC. This can have a negative performance impact due to the lack of database update parallelism and is especially harmful if very long transactions can block shorter transactions which follow them. For instance, consider a large transaction that updates 10,000 records and a following short transaction that commits just after the first transaction commits. It may take several minutes for the first transaction to be applied to the target database once it commits at the source database. The target commit of the following small transaction will be delayed these several minutes as it waits for the large transaction to complete. Furthermore, though processing load was typically level at the source system, the target system will suddenly experience a peak processing load when the large transaction is unleashed. This makes tuning the target system extremely difficult.

As noted above, the DOC can either be on the source system or on the target system. The advantage of placing it on the source system is that it can provide transaction retention should the network or the target system fail. However, in this case, not only will the target system be heavily loaded during the large transaction replay, but so will the communication line. Furthermore, following transactions are now delayed not just by the processing times of the large transaction's updates, but also by the transmission time of these updates across the communication channel. The net result of all of these delays is an increase in the amount of data that may be lost should the source system fail.

Multi-Threaded Replication Engine

Both of the above architectures have a performance limitation because they are single-threaded. That means that there is only one replication channel over which all database changes must flow. In addition, if a DOC is used, the transactions themselves are generally applied one at a time. Without the DOC, the applying of transactions to the target database can be done in such a way that several transactions are active at a time, just as they were at the source system.

Some replication systems today use multiple threads to significantly enhance the capacity of the replication channel. This is shown in Figure 10-5c. In such an architecture, there are multiple replication channels over which changes can be sent from one or more Change Queues through (optionally) one or more DOCs or equivalent to the target database. Of course, if nothing is done to correct it, there is no semblance of natural flow at the target database. Changes are made in arbitrary order, and the database will typically not be in a consistent state.

Multi-threaded architectures are typically used for unidirectional data replication for system backup. Should the source system fail, the target replication engine must first go through the database and make it consistent. This involves completing any transactions that it can and then aborting all transactions that have not completed. Since the unstructured update of the target database usually means that transaction boundaries are also not ordered properly, this can be a complex and timely process.

Dr. Bill Highleyman, Paul J. Holenstein, and Dr. Bruce Holenstein

This architecture, of course, cannot be used in an active/active environment. Often, however, the target system is used for query and reporting purposes. It is important that the users understand that they are working with an inconsistent database and that the results of a query or report may be meaningless.

Multi-threaded replication engines can provide significantly improved performance, however, and will result in a consistent target database if care is taken to re-serialize change events into natural flow order prior to applying them to the target database. The remainder of this chapter discusses techniques for doing just this.

Summary

Maintaining a consistent and uncorrupted database at all nodes in a replicated system network is of paramount importance if the distributed applications are to work properly. When using asynchronous replication, database consistency and integrity can be compromised in several ways:

a) Data collisions may give different views of the database at different nodes, thus corrupting the database.

b) If related changes are not applied as an atomic entity, the database may not be in a consistent state until all of the changes have been applied.

c) If changes or transactions are made in a different order at different nodes, the value of a data item may vary at different nodes and may lead to a corrupted database.

d) If changes or transactions are made in a different order at different nodes, then referential integrity may be violated.

In the simple model of Figure 10-2, the data replication engine preserves transaction boundaries and natural flow so that corruption types b), c), and d) will not occur (data collisions may still occur if replication is asynchronous, though there are techniques to detect and resolve data collisions, as discussed in Chapter 3, *"Asynchronous Replication"*).

However, more complex data replication models do not necessarily preserve either natural flow or transaction boundaries and may be subject to any of these corruption mechanisms. In those cases which are important to the application, there must be a mechanism to serialize events and/or transactions prior to their being applied to the target database. A good bit of the following discussion deals with techniques for event and transaction serialization.

In the following discussions, it is assumed that the data replication engine is asynchronous or weakly synchronous. The focus is on the prevention of database corruption in complex data replication engines due to improper ordering of changes or transactions (corruption types b), c), and d)). These are the examples of the corruptions that can occur when natural flow is lost.

Multi-Threading for Performance

Transaction processing systems today can have transaction rates which are far in excess of what Figure 10-2's simple data replication engine can handle. One solution to this is to multi-thread the replication engine. *Multi-threading* means that multiple parallel paths are provided for data item modifications or transactions to flow from the source database to the target database. In some software models, a thread manager is provided to manage several threads within one process. In other models, threads are implemented as similar processes running in parallel.

Extending the simple model of Figure 10-2, Figure 10-6a shows the multi-threading opportunities for a data replication engine. One can provide parallel Extractors, parallel communication channels, and/or parallel Appliers.

In addition, outside of the replication engine, multiple application programs may be updating multiple databases, each with multiple audit trails, DOCS, or Change Logs resident on different nodes; and the replication engine may be updating multiple target databases.

As shown in Figure 10-6b, in such a multi-threaded environment, transactions and perhaps even the changes within transactions may be flowing across multiple independent paths from source to target. Unless care is taken, there is no control as to the order in which transactions or their changes will arrive at the target system and be applied to the target database, thus leading to the previously described potential database corruption when natural flow is abandoned.

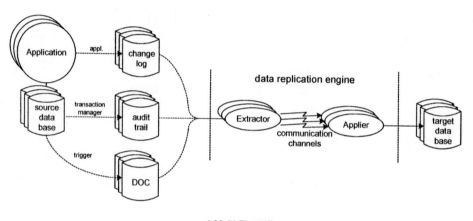

a) Multi-Threading

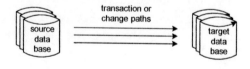

b) Multiple Paths

**Multi-Threaded Data Replication Engine
Figure 10-6**

In the following sections, various multi-threaded architectures are described along with methods to enforce natural flow.

Multi-Threaded Extractor

The Extractor can be made up of multiple extraction threads or multiple extraction processes. There are many possible architectures, in some cases driven by the source database architecture.

In any case, there must be some algorithm to ensure that change events are applied to the target database in the correct order.[41] This algorithm can be a set of rules for what is sent over each Extractor thread, it can be a facility that allows the Extractors to coordinate their activities with each other, or it can be a facility that serializes all updates within a transaction at the target system. Extractor-based algorithms include the following:

Rules-Based Extractor Assignment

Rules-based algorithms include assigning each file or table to a specific Extractor. If there are files or tables that have a referential integrity relationship with each other, then the entire group should be processed by a particular Extractor. In this way, any modification sequence that may have a referential integrity impact is sent in natural flow order.

This, of course, implies that, once an event enters an Extractor, it remains in that thread throughout the replication process until it is applied to the target database. That is, there is a plurality of Extractor/Communication/Applier threads, or *end-to-end threads*, as shown in Figure 10-7. Each handles a set of related files or tables. Since each thread handles all events which are related, then each may apply its events independently of the other threads.

---

[41] The requirement for natural flow can be relaxed in some cases such as restart scenarios, providing that the brief period of referential integrity relaxation is corrected at the end.

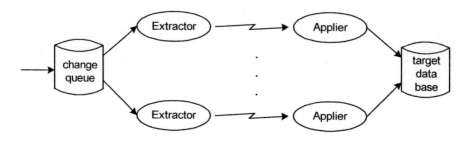

**Separate End-to-End Threads
Figure 10-7**

As discussed later, if end-to-end threads are not used and transactions can span threads, there may be no guarantee of event order at the target system. Therefore, a re-serializer may be needed prior to applying changes to the target database.

If several Extractors are involved in a transaction, the problem exists of how the begin/end commands are replicated since they are not associated with a file or table. A solution to this may be characterized as "expected ends" (Figure 10-8a). In principle, some Extractor is going to find a begin transaction command and will send it to the target system. (In some systems, a begin transaction command is implicit in the first modification for a new transaction.) This might be a Master Extractor which is charged with the responsibility for sending begin/ends. It might be the first Extractor to find the begin if multiple Extractors are sharing the same Change Queue. If there are multiple Change Queues each serviced by its own Extractor, it might be the Extractor that finds the begin transaction command. In any event, there must be a mechanism that guarantees that the begin transaction will be applied once and only once.

Breaking the Availability Barrier

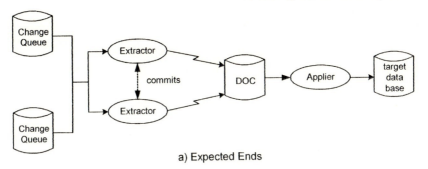

a) Expected Ends

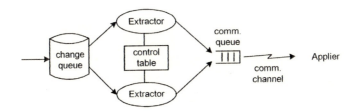

b) Extractors Coordinating through a Control Table

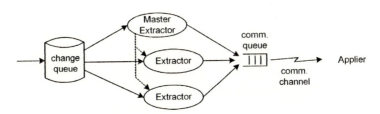

c) Extractors Controlled by a Master Extractor

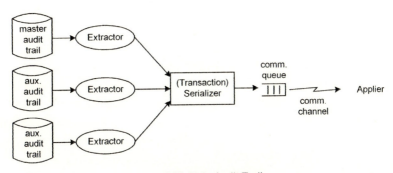

d) Multiple Audit Trails

**Multi-Threaded Extractors
Figure 10-8**

Likewise, some Extractor will find the commit transaction command. It is not sufficient for that Extractor to simply send the commit to the target system and for the target system to take a commit action. All modifications associated with that transaction may not have reached the target system if they are traveling over other threads. Rather, a commit token of some sort (perhaps the commit command itself or some representation of it, such as its location in the Change Queue) is sent over all Extractor threads (or at least over those threads that are involved in the transaction). Each Extractor thread must insert the commit token in proper time or event sequence into its stream, at least to the extent that the commit is not inserted ahead of other related events that occurred earlier. When the commit token is received at the target system from all of the Extractor threads (or at least those involved in the transaction), then the transaction can be committed since it is now guaranteed that all modifications have been received from all involved Extractor threads. In effect, the data replication engine can run at the speed of the slowest thread.

So far as the data modifications are concerned, the order of arrival at the target database is not guaranteed because the modifications may travel over separate threads. If all modifications for the same file are sent over the same thread, thus guaranteeing their order, they can be applied as soon as they are received so long as inter-transaction referential integrity is not an issue. The last thread to receive the commit token will be responsible for committing the transaction (either itself or by notifying the transaction owner – the thread that began the transaction).

However, if the database manager checks referential integrity on each modification event, then an event sequence which causes a referential integrity violation may cause the database manager to reject the transaction. If this is a concern, then data modifications must be properly re-serialized at the Applier, as we discuss later (Figure 10-8a shows a DOC at the target node acting as a re-serializer).

Breaking the Availability Barrier

Separate end-to-end threads guarantee the natural flow of changes within a file or table and between related files and tables. However, they do not guarantee the natural flow of transactions. Therefore, their use alone is limited to those applications in which there are no inter-transaction dependencies.

If inter-transaction order is important, and if the related transactions can travel over different threads, then a re-serializer is required at the target node to ensure proper transaction order. Re-serialization at the target node is discussed later under Multi-Threaded Applier.

Inter-Extractor Coordination

Some additional examples of Extractor coordination mechanisms are shown in Figure 10-8.

Figure 10-8b shows multiple Extractors directly reading the Change Queue. Some means by which to coordinate the Extractors is required so that it is guaranteed that each transaction is claimed once and only once by an Extractor.

One way to assign transactions to the Extractors is via a Control Table. This Control Table may be maintained in memory to improve performance, or it may be disk-resident to aid in recovery (following a source node or network failure, for example, a durable copy of the Control Table can indicate which transactions have been successfully replicated or to what point in the Change Queue the Extractor had progressed).

Change modifications must be sent to the Extractors by some method. One method is for all Extractors to read the Change Queue for all changes. Each is looking for a begin transaction command (which may be implicit in the first change event). When it finds one, it will lock the Control Table and check to see if this transaction has been claimed by another Extractor (if the Control Table is already locked, the Extractor will wait until it can acquire the lock). If the transaction has already been claimed, then the Control Table is unlocked; and the Extractor continues its scan of the Change Queue.

If the transaction has not yet been claimed, the Extractor will enter the transaction id in the Control Table, unlock the Control Table, and send the begin transaction command to its Applier, which will begin a transaction at the target database. (An equivalent to the Control Table is to note in the Change Queue that the transaction has been claimed.)

The Extractor will then continue to read changes associated with this transaction as well as the associated end transaction command and will send these to its associated Applier to be applied to the target database.

The Extractor will then return to the point in the Change Queue where it found the begin transaction command and will continue scanning.

Optionally, the Applier may return an acknowledgement to the Extractor that the transaction has been successfully applied. The Extractor may mark the transaction's entry in the Control Table or Change Queue to reflect this or alternatively delete the entry from the Control Table or Change Queue.

An alternate to the Control Table is for each Extractor to take every nth transaction, where n is the number of Extractors. For instance, if there are three Extractors, Extractor 1 will take transactions 1, 4, 7, ..., Extractor 2 will take transactions 2, 5, 8, ... and so on.

This description is for the simple case of each Extractor handling just one transaction at a time. In many implementations, each Extractor may be handling multiple overlapping transactions at a time. The techniques above are also applicable to this case. Certain problems related to a thread handling multiple simultaneous transactions are discussed later.

There are other ways in which transactions may be allocated to threads. For instance, transactions may be allocated to threads based on which files or tables they change, as described above. Alternatively, all transactions that may have inter-transaction consistency or referential integrity issues may be sent down the same thread.

Figure 10-8c shows another arrangement in which one Extractor is the Master Extractor and the other Extractors are slaves to it. The Master Extractor reads entries from the Change Queue and looks for a begin transaction command. When it finds one, it will assign this transaction either to itself or to one of its idle slave Extractors. The assigned Extractor will then read the changes and end command for that transaction and will queue them to the communication channel for transmission to the target system.

This coordination technique of using a Master Extractor is also applicable to end-to-end threads, as shown in Figure 10-7.

In some cases, the database will distribute its change records over multiple disk volumes to improve performance. In one such implementation, there is a Master Audit Trail disk volume and one or more Auxiliary Audit Trail disk volumes (Figure 10-8d). The Master Audit Trail contains the begin/end commands, possibly certain change records, and pointers to all other change records for each transaction, thus preserving natural flow. The change records are distributed across the Auxiliary Audit Trails.

An Extractor is provided to read one or more audit trail volumes. Each Extractor sends its data to a common Serializer which maintains in memory the begin/end command for a transaction, pointers to its change records, and the change records as read from the Master Audit Trail and the Auxiliary Audit Trails. It is the job of the Serializer to queue intra-transaction data to the communication channel in proper order (begin, changes, end). The Serializer may also be responsible for queuing the transactions themselves to the communication channel in natural flow order. In this case, through the use of a Transaction Serializer, all transaction data will be sent over the communication channel in natural flow order.

A Transaction Serializer can also be used with the architectures of Figures 10-8b, 10-8c, and 10-7. Without a Transaction Serializer, these architectures guarantee natural flow within a transaction but not between transactions. With a Transaction Serializer, any multi-threaded Extractor architecture will provide natural flow both within transactions and between transactions.

If the communication channel is single-threaded, natural flow is preserved at the input to the Applier. If the Applier is single-threaded, natural flow is preserved at the target database.

Multi-Threaded Communication Channel

If the transaction rate is so high that it will overload a single communication channel, then the load can be split across multiple communication channels. There are several ways to do this.

One way is for the Extractors to pass their transaction data to a common communication queue which is read by all communication channels, as shown in Figure 10- 9a. However, any semblance of natural flow even within a transaction will be lost as changes will be delivered in unpredictable order to the Appliers. Unless there is a higher communication level that can guarantee that messages will be delivered in proper order, a Serializer will be required at the target system. This is described later.

Alternatively, each Extractor can feed its own communication line, as shown in Figure 10-9b. This will protect the intra-transaction natural flow as long as the Extractors have provided intra-transaction natural flow. However, in this architecture, neither the Extractors nor the communication channels support inter-transaction natural flow.

If transaction natural flow must be delivered to the target node because there is no re-serialization capability in the target node, then multiple communication channels cannot be used unless a higher level is provided to place messages into the proper order. Even if a Transaction Serializer is used to serialize transactions feed to the communication queue by the Extractors, as shown in Figure 10-9c, all serialization will be lost when transactions are sent over separate communication channels that did not have a reordering capability at the receiving end.

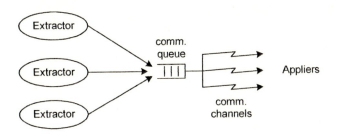

a) Common Communication Queue

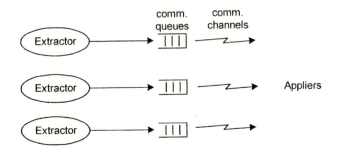

b) Independent Communication Channels

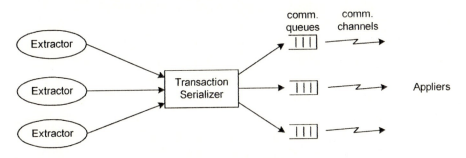

c) Serializer is Ineffective

**Multi-Threaded Communication Channel
Figure 10-9**

Unless the communication channels as a group can guarantee delivery in the proper order, the only way that transaction natural flow can be guaranteed to be delivered to the target system is to use a single communication channel between the source and target system. If only intra-transaction natural flow is required, then multiple communication channels can be used by associating each with a specific Extractor, providing that each transaction is sent by a single Extractor.

Multi-Threaded Applier

To obtain the required performance, the Appliers may also be multi-threaded. In the case shown in Figure 10-10a, transactions with changes in natural flow order are received by a Router, which routes each transaction to one and only one Applier (though each Applier can be handling several interleaved transactions simultaneously). Each Applier starts its transaction, applies the transaction changes, and then commits the transaction (or aborts it if the source system has aborted it).

Suspend on Commit

In the architecture of Figure 10-10a, each transaction is guaranteed to have its changes applied in natural flow order. However, additional precautions must be taken by the Router if transaction order is to be preserved. To accomplish this, the Router will still send multiple transactions to multiple Appliers, one or more transactions per Applier. However, the Router must guarantee the correct commit order via some algorithm. One such algorithm is for the Router to serialize all change events in its own memory space and to distribute properly serialized begin transaction commands and updates to its Appliers until it reaches a commit token (which could be a set of commit tokens received over all threads involved in a transaction). At this point, it will suspend the routing of begin and change information and will just send a commit command to the Applier managing that transaction. It will wait for that Applier to acknowledge that it has completed the commit operation and then will continue sending further begin and update messages to its Appliers. In this way, transaction order is guaranteed.

Breaking the Availability Barrier

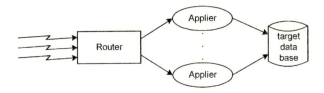

a) Simple Router

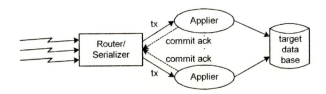

b) Router with Referential Integrity

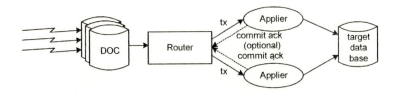

c) Routing from a DOC

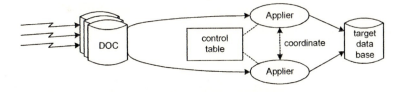

d) Inter- Applier Coordination

**Multi-Threaded Appliers
Figure 10-10**

Alternatively, the Appliers, via a common table or some other synchronizing means, will coordinate their updates based on some ordering criteria such as a sequence number or a date/time stamp.

Appliers Coordinate Commits

Requiring that the Router hold up the distribution of begins and changes while a commit is taking place slows down the replication process. An alternative strategy is to simply let the Router distribute begin, change, and commit events as it receives them. It then assigns each transaction to a specific Applier, as described above. However, all Appliers apply their transactions as soon as they have them and coordinate only their commits. An Applier will hold up a commit if there are earlier commits outstanding. If two transactions attempt to update the same data item out of order, then the Applier with the earlier commit will not be able to lock the data item; and the Applier with the later commit will not be able to commit and unlock the item. Consequently, a deadlock occurs.

This deadlock is easily resolved by aborting the newer transaction (i.e., that transaction with the later commit) that is holding the data item lock out of order (or at least requiring that the Applier improperly holding the lock release the lock). This will let the earlier transaction complete, and the newer transaction can then be retried by the Applier that had to back off. By allowing such deadlocks, all Appliers may be running at near full speed; and replication capacity may be greatly enhanced. Even if there are a lot of deadlocks, this approach will often yield higher throughputs than the earlier techniques, which allow only one transaction commit at a time.

Multiple communication lines can be used to send change data to the Router since the Router can serialize transactions. In fact, though the above description has assumed that changes within a transaction will be received in natural flow order, the Router can be designed to provide complete serialization, both intra-transaction and inter-transaction.

Using a DOC

In some cases, it will be advantageous to write all incoming transactions to an intermediate disk-based or memory-resident DOC. This is useful, for instance, if there is no transaction manager at the target database. Using a DOC allows aborted transactions to be filtered out and not applied to the target database (a Router, if used, can also provide the same function). A DOC will also provide queuing storage to accommodate peak change rates that may overwhelm the Appliers.

If a DOC is provided, there are several architectures to read changes and then to apply them with multi-threaded Appliers to the target database. As shown in Figure 10-10c, one way is to provide a Router to read the DOC. In this case, the Router is much simpler in that the DOC can provide both intra-transaction and inter-transaction serialization functions via its indices. The Router can request the next transaction in commit order and can assign it to an Applier. It can then read the changes for that transaction in order or request the Applier to do so.

If transaction natural flow is to be preserved, then any of the procedures described earlier can be used. If the Router is reading the DOC, it can suspend distribution of updates when it sends a commit to an Applier until the Applier acknowledges that it has completed the commit. Alternatively, all Appliers can be simultaneously reading the DOC and updating the database, pausing only to coordinate their commits.

Alternatively, in a manner similar to that described for multi-threaded Extractors with reference to Figure 10-8b, multiple Appliers can coordinate with others, as shown in Figure 10-10d. All Appliers read the DOC and use a Control Table to allocate transactions. However, transaction commits are coordinated by communication between the Appliers or alternatively via the Control Table. In this case, a virtual Control Table can be implemented via DOC keys. By properly structuring the keys to the transaction commits stored in the DOC, each Applier can read the next free transaction in commit sequence.

If the Appliers apply modifications as they receive them, and if either they or the Router coordinate the commits, then the techniques described above guarantee that transactions are applied in natural flow order; unfortunately, modifications may not be. As a result, deadlocks may occur. However, if each Applier holds its modifications until the previous transaction has committed, then natural flow order of transactions and modifications is guaranteed. The consequence of this is reduced performance, since only one transaction at a time is being applied to the target database.

It must be noted that a DOC may provide many of the functions attributed to other components in the descriptions above by making judicious use of indices into the DOC. These functions include:

- routing to the Appliers (in this case, a Router is not required).

- serializing events to the Appliers (in this case, a Serializer is not required).

- coordinating the Appliers (in this case, a Control Table is not required).

These functions can be provided by the DOC whether it is disk-based or memory-resident.

Exceeding Transaction Count Limits

In the systems which we have considered, many applications are active simultaneously and, in fact, may themselves be multi-threaded. All application threads may be generating transactions independently. Although each application thread is generally only processing one transaction at a time, the multiplicity of application threads means that at any point in time there are many transactions that are in progress. This mix of transaction activity is represented in the Change Queue (whether it be a change log, an audit trail, or a DOC) by intermixed entries for several transactions.

Breaking the Availability Barrier

Each Applier must manage each of its transactions as that transaction is applied to the database. It must begin the transaction, apply the transaction's modifications to the target database, and then commit (or abort) the transaction. During the time that the transaction is open, the Applier *owns* the transaction.

In many systems, an Applier may own several transactions simultaneously (typically, one per thread). The requirement for natural flow means that the Applier must execute these transactions in the same or similar order as they had been executed at the source system. Therefore, the Applier must be able to accept all new transactions as they are assigned to it.

However, in many systems there is a limit as to how many transactions a process may own. If the Applier reaches its transaction limit, it cannot process the next begin transaction command, which means that it cannot process the rest of the changes for its currently open transactions and still maintain natural flow order. In effect, the replication process is blocked; and replication halts.

There are several ways to resolve this problem.

Multiple Appliers

The use of a multi-threaded Applier was described in reference to Figure 10-10. Several such Appliers may be provided to share the transaction load. To guarantee natural flow order, the actions of the Appliers are coordinated through means such as a Router/Serializer or a Control Table.

With this technique, should the number of concurrent transactions become too large for the current set of Appliers, then additional Appliers may be spawned. When the transaction load diminishes, excess Appliers may be terminated.

One problem with a multiple Appliers is that transaction commits must be serialized as described above if natural flow is to be preserved, thus posing a potentially significant performance problem. That is, by coordinating transactions with each other, the Appliers guarantee that transactions will be executed in natural flow order but may have to pause often while earlier commits are applied.

As a consequence, the application of transactions is not fully multi-threaded; and the resultant set of multiple Appliers may not provide the performance enhancement that is expected.

Partial Transactions

Another approach to the problem of too many transactions for a single-threaded Applier is to limit the number of concurrent transactions that the Applier is handling by prematurely committing outstanding transactions if need be.

More specifically, if the Applier reaches its concurrent transaction limit, it will choose one or more existing open transactions and will commit them prematurely, thus freeing up slots for following transactions. Later, should another modification for the prematurely committed transaction arrive at the Applier, then the Applier will start a new transaction for the recently received modification. If the transaction limit is once again reached, the above process is repeated.

As a consequence, during periods of peak activity, a transaction might be broken up into two or more sub-transactions. During this period, the database may exhibit inconsistency and referential integrity violations. However, all transactions and data modifications are applied in natural flow order; and the database will return to a correct state when the transaction rate diminishes and all partially committed transactions have been fully committed.

Unless the Applier is being driven by a mechanism such as a DOC to filter out aborted transactions, some of these partially committed transactions may eventually abort. The replication engine will not be able to use the target system's abort facility to abort the previously committed partial portions of the transaction. Therefore, it is important that the replication engine replicate undo events or before images for aborted transactions. An undo event for an update is the before image for that data item. An undo event for an insert is a delete, and an undo event for a delete is an insert. These undo events should be replayed at the target system in natural flow order.

Adaptive Replication Engine

One can combine all if these approaches to balance performance and complexity by having the replication engine be adaptive. One such architecture is described below.

During low traffic times, a single Applier is used between the source database and the target database.

If the Extractor discovers that its Applier is approaching the transaction limit, it can spawn another Applier at the target system and establish a connection with it. It can then distribute transactions between its two Appliers. If these Appliers start to become full, the Extractor can spawn another Applier, and so on.

If the number of Appliers reaches a system limit, then the Extractor can command the Appliers to do partial commits to limit the number of outstanding transactions in order to remain within the system limits. During this time, the target database will not be in a consistent state to a viewer; but no transactions will be lost.

As the traffic peak diminishes to an appropriate level, the Appliers can be commanded to return to normal transaction committing. After a short period, the target database will return to a consistent state. As traffic further diminishes, excess Appliers can be terminated until the replication engine returns to its basic single-threaded state.

Dr. Bill Highleyman, Paul J. Holenstein, and Dr. Bruce Holenstein

Resolving Deadlocks

Deadlocks With An Application

In attempting to lock data items for modification, there is a chance that the replication engine may deadlock with an application process. This can be resolved either by the replication engine or by the application timing out, backing off, releasing all colliding locks, and retrying the transaction after a short delay. By this time, the other entity should have finished its transaction and have released its locks. If not, the replication engine or the application can back off again. Alternatively, the replication engine can terminate the offending application.

If there is a deadlock with an application, this means that both the application and the replication engine were trying to update the same data item. Thus, a data collision has most likely occurred. However, data collisions are expected in an asynchronous active/active replication environment; and this is just another case of a data collision.

Deadlocks With Another Applier

However, it is also possible that two Applier threads may deadlock with each other. This may occur for a variety of reasons. For example, if the target database schema does not match the source database schema, and hence the transaction locking profiles are different, then a source-locking protocol that is meant to avoid deadlocks may not be applicable to the target system. In this case, backing off may not solve the problem since the same situation may reoccur. This problem can be resolved by committing partial transactions, as described previously. In effect, one or both threads will commit the partial transaction that they currently are managing. Each thread that does this then opens a new transaction to complete its respective transaction. Each new transaction can then modify its deadlocked data item, which the other transaction had already modified and unlocked in the partial transaction.

Alternatively, the Applier with the newer transaction commit-wise can back off by releasing its conflicting lock(s) to let the older transaction complete.

Deadlocks With Another Transaction

There are two different ways that a database manager might audit changes to its database:

a) <u>Logical Level Audit</u> – Each logical modification to the database is logged to the audit file. Logical <u>modifications</u> include inserts, updates, and deletes. Each is replicated to the target database by the replication engine.

b) <u>Base Level Audit</u> – Each physical disk operation is logged to the audit file. The physical operations are replicated by the replication engine to the target system, where they are applied to the target database either as physical or logical events.

Base level auditing presents a unique problem to the replication engine because each replicated event may not represent a complete logical operation. For instance, a logical insert to a file or table with multiple indices will be replicated as one update to the base file or table and one update to each of the index files. The replication engine may replicate all such updates to their respective target files. Alternatively, the replicator may apply the base file or table update as a logical update (which will update all target index files whether the schema is the same or not) and will then ignore the index file updates.

Certain confusing locking conditions can occur within a single Applier when updates to a file or table with multiple unique indices are replicated from an audit file that uses base level auditing via a logical replication engine. These can be managed by one of several algorithms:

a) Accumulate Index Inserts

When a base table insert is found, hold it until all significant index inserts (for instance, all unique index inserts) are also received. At this point, the base table logical update can be safely applied.

b) Set-Aside Queue

If a database access times out on a lock, place it in a first-in, first-out Set-Aside Queue. Append to the end of the queue any further modifications to the affected table or table partition, even if they are being applied by other transactions. Also, place further events for this transaction and optionally other events for those other transactions in the queue if the preservation of intra-transaction natural flow is desired. That is, once an event is on the queue, always append related events for the same table or table partition (and optionally for all participating transactions) to the queue.

Upon the occurrence of certain significant events, attempt to replay the Set-Aside Queue. Replay all events in the queue until a lock is found (or the queue is emptied). Significant events may include:

- commits/aborts (locks may have been freed up).

- time interval expiration (locks held by other applications may have been freed up).

- other (e.g., a certain number of events or transactions have been processed).

With this procedure, intra-transaction natural flow is preserved as is inter-transaction natural flow so far as each table is concerned. Inter-table natural flow is not preserved. This can cause some referential integrity violations such as the improper creation order of parent/child records or rows. It can be mitigated by having the replicator consider such events to be related in this algorithm.

A third algorithm which can be applied to this problem is:

c) <u>Asynchronous Database Access</u>

Use asynchronous file access to access the database. In this way, a lock will not hold up the replicator; and the database accesses to follow can be initiated in natural flow order.

In effect, asynchronous database access allows all database accesses to be initiated in natural flow order. These accesses will be executed in natural flow order so long as the database preserves initiation order during execution.

There are many ways to implement asynchronous database access. They include asynchronous calls, a separate thread for each transaction, or parallel Appliers.

Summary

The previous discussions have described various architectures for multi-threading Extractors, communication channels, and Appliers. Many of these architectures, including permutations, can be used in conjunction with each other to create powerful replication engines. Some permutations and combinations are shown in Figure 10-11. Each has different performance and serialization characteristics.

Figure 10-11a shows a single Extractor also acting as a Router. It reads the Change Queue and sends each transaction to a specific Applier on the target system. The configuration preserves the natural flow of changes within a transaction, but does not preserve transaction order. The target system is multi-threaded, but the source Extractor is single-threaded.

However, the architecture of Figure 10-11a can be used to also guarantee transaction order. To do this, the Extractor will suspend its activity following the sending of a commit command to an Applier until that Applier has acknowledged that it has completed the commit.

Alternatively, the Appliers can coordinate their commits among themselves. Both of these strategies have been described earlier.

In Figure 10-11b, multiple Extractors read one or more Change Queues and send all changes to a Serializer. The Serializer delivers a change flow to the communication channel in full natural flow order for both intra-transaction changes and for the transactions themselves. A single communication channel and a single Applier ensure the natural flow of change data to the target database. The source system is multi-threaded, but the target system is single-threaded.

Figure 10-11c shows multi-threaded Extractors sending changes over a single communication channel to a target Router. The Router provides serialization services and routes each transaction to an appropriate Applier. Commits are coordinated via one of the algorithms previously described. Changes and transactions are applied to the target database in natural flow order, and both source and target systems are multi-threaded. This architecture will also work with multiple communication lines. With a single communication line, the natural flow of changes within a transaction can be controlled by the Extractors. However, if multiple communication lines feeding from a common communication queue are used, then proper change order must be provided by a Router/Serializer on the target node unless a higher communication level is provided to guarantee message order upon delivery.

Figure 10-11d shows the configuration of Figure 10-11c but with a DOC at the target system receiving and storing all changes before they are applied to the target database. In this case, the DOC can provide all serialization functions; and the Router is much simpler. It needs only to route transactions and sequence commits. In fact, the functions of the Router can be implemented via the key structure of the DOC coupled with proper coordination between the Appliers.

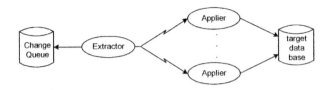

a) Single Extractor, Multi-Threaded Applier

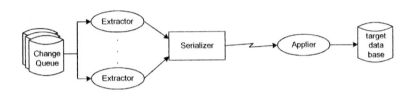

b) Multi-Threaded Extractor, Single Applier

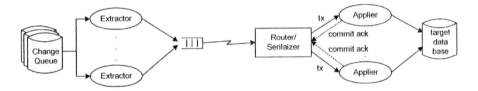

c) Multi-Threaded Extractor, Multi-Threaded Applier

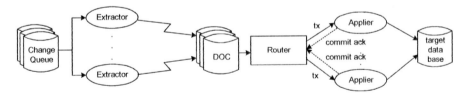

d) Multi-Threaded Extractors, Comm Channels, and Appliers

Some Data Replication Engine Configurations
Figure 10-11

Figure 10-11 shows only some permutations and combinations of replication engine components. For instance, a combined Extractor/Applier may be resident on the source system and "push" changes to the target database using RPCs (remote procedure calls) or some equivalent mechanism. Likewise, a combined Extractor/Applier may reside on the target system and "pull" changes from the source system via RPCs or an equivalent mechanism.

Furthermore, the interaction between the Extractors and Appliers may take different forms:

- In the simplest case, an Extractor sends just one update or begin/end command at a time. It waits for an acknowledgement from its Applier before it sends the next one. If there is only one Extractor, then this guarantees natural flow order but with a serious performance penalty.

- An Extractor can send begin commands and updates as fast as it can but pause for acknowledgements on commits.

- an Extractor can send all events as fast as it can and let the Appliers coordinate their commits.

- If tránsaction order is not important, Extractors and Appliers can both operate at full speed.

From a performance viewpoint, the configuration shown in Figure 10-11c is the best choice because it is fully multi-threaded and has no queuing points. Multiple communication lines can be provided. The Extractors can operate at full speed. The Appliers can also proceed at full speed, pausing only to coordinate their commits (and possibly to resolve deadlocks).

The point is that there are many ways to configure a data replication engine to achieve an application's performance and database consistency requirements. But attention must be paid to natural flow if the database is to remain consistent and uncorrupted. This generally means that:

> **Rule 35:** A *serializing facility that will restore natural flow is required following all data replication threads and before the*

target database in order to guarantee that the database will remain consistent and uncorrupted.

The one exception to this rule is when there is no inter-transaction relationship to be maintained. If the application is such that transactions can be applied in any order, then it is only necessary to serialize modifications within a transaction. This can be done by ensuring that all modifications associated with a given file, table, or file/table partition are always sent over the same thread.

Chapter 11 - Failover Faults

When a redundant system fails, it is often necessary not only to repair the requisite number of failed subsystems, but then to recover the system. Recovery may involve cold loading the operating system, restarting applications and networks, and perhaps recovering the database. Recovery may require several hours before the system becomes available.

In Chapter 5, *"The Facts of Life,"* we intuitively derived certain effects of repair and recovery. Chapter 11 presents several repair/recovery models for redundant systems. These models express the probability of failure of such systems and consider not only hardware and software failures that require recovery following repair but also the impact of failover faults that cause an outage following a single failure. The models also consider a variety of repair strategies.

This chapter first intuitively derives approximations for failure probabilities for each case. The approximations are supported in Appendix 3 by a more exact analysis using Markov models. Appendix 3 is an extension of the work of Dr. Alan Wood.[42]

The following important insight is derived:

Once one subsystem has failed, failover faults cause the system to behave as if it is comprised of (n-1) subsystems, each with decreased availability.

The effective availability and repair time for each model is tabulated. It is demonstrated that even a small probability of a failover fault can have a significant impact on system availability.

---

[42] Wood, A.; *"Availability Modeling,"* Circuits and Devices; May, 1994.

Dr. Bill Highleyman, Paul J. Holenstein, and Dr. Bruce Holenstein

Failover Faults

We consider redundant systems comprising n identical subsystems. The system is designed to tolerate the failure of any single subsystem. However, the failure of two subsystems will result in a system outage.

In order to return the system to service following an outage due to a dual subsystem failure, at least one of the subsystems must be repaired. Following this, the system may or may not need to go through a recovery process before it can be returned to service.

This scenario is dependent on each subsystem having a backup. When a subsystem fails, its functions are assumed by a surviving subsystem. We call this process *failover*. One complication is that the failover may fail (a *failover fault*), caused perhaps by an error in the software failover logic or by an erroneous and fatal operator action. In any event, the system suffers an outage following a single failure even though a perfectly good spare is available.

Following an outage due to a failover fault, the system usually does not need to be repaired since only one subsystem has failed and since the system is designed to operate properly when one subsystem is down. However, the system probably does need to be recovered.

The models which follow consider several failover fault scenarios:

(1) <u>No Failover Faults</u> - The system is not subject to failover faults.

(2) <u>Failover Fault Requiring Subsystem Repair</u> - Given a failover fault, the failed subsystem must be repaired; and the system is returned to service with no recovery required following this repair.

(3) <u>Failover Fault Requiring System Recovery</u> - Any outage, whether caused by a dual subsystem failure or by a failover fault, requires a system recovery. System recovery

time is different than subsystem repair time. This is the case that most closely reflects the real world.

Repair Strategies

The following models consider three repair strategies:

a) <u>Parallel Repair</u> – When multiple subsystems have failed, repair proceeds independently via separate service technicians working on each subsystem. The return to service of one subsystem is independent of the others.

b) <u>Sequential Repair</u> – Only one service person is available, and he works on one subsystem at a time. Therefore, one subsystem is repaired and returned to service; and then the next subsystem is repaired.

c) <u>Simultaneous Repair</u> – All subsystems are repaired simultaneously. A possible scenario for this is that following a multiple failure, a service person is dispatched. He diagnoses the problems, leaves to fetch spare parts, and upon his return repairs all subsystems and returns them to service at the same time. Alternatively, a spare system may be swapped in.

No Failover Faults

Let a be the availability of a subsystem with a mean time before failure *mtbf* and a mean time to repair *mtr*. a is the probability that the subsystem is operational (its availability):

$$a = \frac{mtbf}{mtbf + mtr} \approx 1 - \frac{mtr}{mtbf} \qquad (11\text{-}1a)$$

where the approximation depends upon $mtr \ll mtbf$.

The probability f of a subsystem outage is

$$f = 1 - a \approx \frac{mtr}{mtbf} \qquad (11\text{-}1b)$$

A subsystem will be out of service with a probability of $(1-a)$. Assuming that failures and repairs are independent (i.e., repairs are done in parallel) then the probability that the system will be unavailable is the probability that two subsystems are unavailable (we ignore the cases of more than two subsystems being out of service as this is relatively unlikely).

The probability that a particular pair of subsystems will be simultaneously unavailable is $(1-a)^2$. If there are n subsystems, there are $n(n-1)/2$ combinations of subsystems that can fail and will cause a system outage. Thus, the probability, F, that the system will be in failure is

$$F \approx \frac{n(n-1)}{2}(1-a)^2 \quad \text{(parallel repair)} \qquad (11\text{-}2)$$

This relationship has assumed that subsystem repairs are independent and is the case for parallel repairs. If there are two subsystems being repaired, each with a mean time to repair of *mtr*, then it can be shown[2] that the average time for the first subsystem to be repaired is *mtr*/2. Once the first subsystem has been repaired, then the system can be restored to service.

However, if the two systems are being repaired sequentially, then the first subsystem will become available in an average time of *mtr*. As can be seen from Equation (11-1b), failure probability is proportional to repair time. Thus, the system will be down twice as long as with parallel repairs and

$$F \approx n(n-1)(1-a)^2 \quad \text{(sequential repair)} \qquad (11\text{-}3a)$$

---

[2] If $(s+1)$ subsystems are being repaired independently, and if each requires an average time of *mtr* to repair, then they are being repaired at a rate of $(s+1)/mtr$. Thus, the average time to the next repair, which will return the system to service, is $mtr/(s+1)$.

Likewise, if the subsystems are repaired simultaneously, both will be returned to service in an average time *mtr*. This is equivalent to the sequential repair case above:

$$F \approx n(n-1)(1-a)^2 \quad \text{(simultaneous repair)} \quad (11\text{-}3b)$$

Failover Fault Requiring Subsystem Repair

We now consider the case in which a failover will fail with probability p. Following a failover fault, the system requires the repair of the failed subsystem to return it to service. No special recovery is needed.

There are now two outage modes:

(1) Two subsystems have failed.

(2) One subsystem has failed, and failover has failed.

Again, assuming that failures and repairs are independent events, then the probability that both systems are down is given by Equation (11-2). (1-p) of outages will be caused by such dual faults:

$$\text{outage probability due to dual failures} \approx (1-p)\frac{n(n-1)}{2}(1-a)^2$$

The probability that a single subsystem will fail is (1-a). There are n ways in which a system can experience a single subsystem failure. Thus, the probability that a single subsystem will fail is $n(1-a)$. p of all outages are caused by single subsystem failures followed by a failover fault:

$$\text{outage probability due to failover fault} \approx pn(1-a)$$

Recognizing that $p \ll 1$ and that the system failure probability, F, is the sum of these two probabilities, then

$$F \approx \frac{n(n-1)}{2}(1-a)^2 + pn(1-a) \quad \text{(parallel repair)} \quad (11\text{-}4)$$

As pointed out in the previous section, for both sequential and simultaneous repairs, the repair of at least one failed subsystem will be twice that for parallel repairs. However, the repair time following a failover fault is the same for all cases (there is only one failed subsystem). Thus, for these cases,

$$F \approx n(n-1)(1-a)^2 + pn(1-a) \quad \text{(sequential, simultaneous repair)} \quad (11\text{-}5)$$

Failover Fault Requiring System Recovery

The failure scenario which most closely reflects the real world of today is that a system recovery is required following any outage, whether that outage be due to a dual subsystem failure or to a failover fault. For ease of notation, let us define

    r = mean time to repair a subsystem (mtr).
    R = mean time to recover the system.

We consider first the parallel repair case. Again, there are two outage modes:

(1) <u>Two subsystems have failed</u>

The argument made above for dual failures still holds for a dual subsystem failure, except now the repair time is the time to repair the subsystem, $r/2$, plus the time to recover the system, R, giving a total repair time of $r/2+R$ rather than just $r/2$. Since failure probability is proportional to repair time (see Equation (11-1b)), then the failure probability for this mode is increased by a factor of $(r/2+R)/r/2$:

    outage probability due to dual failures

$$\approx \frac{r/2+R}{r/2}(1-p)\frac{n(n-1)}{2}(1-a)^2$$

(2) <u>One subsystem has failed, and failover has failed</u>

Again, the argument that led to Equation (11-4) still holds, except that repair time, r, has been changed to system recovery time, R. Thus, the probability of failure for this mode is modified by a factor of R/r:

$$\text{outage probability due to failover fault} \approx \frac{R}{r}pn(1-a)$$

Recognizing that the system failure probability, F, is the sum of the above two probabilities and that $p \ll 1$, then

$$F \approx \left(\frac{r/2+R}{r/2}\right)\frac{n(n-1)}{2}(1-a)^2 + \frac{R}{r}pn(1-a)$$

(parallel repair) \hfill (11-6)

Using the same argument presented in the previous section, the dual outage repair time will be doubled from $r/2$ to r for sequential or simultaneous subsystem repair; but the recovery time remains the same. Thus, for these cases,

$$F \approx \left(\frac{r+R}{r}\right)n(n-1)(1-a)^2 + \frac{R}{r}pn(1-a)$$

sequential, simultaneous repair) \hfill (11-7)

Failure Model Summary

The above results are summarized in Table 11-1.

Repair	Failover Fault — None	Failover Fault — With Subsystem Repair
Parallel	$\frac{n(n-1)}{2}(1-a)^2$	$\frac{n(n-1)}{2}(1-a)^2 + pn(1-a)$
Sequential	$n(n-1)(1-a)^2$	$n(n-1)(1-a)^2 + pn(1-a)$
Simultaneous	$n(n-1)(1-a)^2$	$n(n-1)(1-a)^2 + pn(1-a)$

Repair	With System Recovery
Parallel	$\left(\dfrac{r/2 + R}{r/2}\right)\dfrac{n(n-1)}{2}(1-a)^2 + \dfrac{R}{r}pn(1-a)$
Sequential	$\left(\dfrac{r+R}{r}\right)n(n-1)(1-a)^2 + \dfrac{R}{r}pn(1-a)$
Simultaneous	$\left(\dfrac{r+R}{r}\right)n(n-1)(1-a)^2 + \dfrac{R}{r}pn(1-a)$

**F
Failure Probability (single spare)
Table 11-1**

An Interpretation

Effective Subsystem Availability

An interesting insight into these relationships can be gained by a little reordering of the terms. Let us use as an example the case for Failover Fault Requiring Recovery with Parallel Repair (Equation (11-6)):

$$F \approx \left(\frac{r/2 + R}{r/2}\right)\frac{n(n-1)}{2}(1-a)^2 + \frac{R}{r}pn(1-a)$$

(parallel repair)　　　　　　　　　　(11-6)

This can be manipulated to be

$$F \approx \left(\frac{r/2+R}{r/2}\right)\frac{n(n-1)}{2}(1-a)\left[1-\left(a-\left(\frac{R}{r/2+R}\right)\frac{p}{n-1}\right)\right] \quad (11\text{-}8)$$

This can be rewritten as

$$F \approx \left(\frac{r/2+R}{r/2}\right)\frac{n(n-1)}{2}(1-a)(1-a') \quad (11\text{-}9)$$

where

$$a' = a - \left(\frac{R}{r/2+R}\right)\frac{p}{n-1} \quad (11\text{-}10)$$

Equation (11-9) has an interesting interpretation. It reflects a system comprising n subsystems. The failure of any two subsystems will cause a system outage (there are $n(n-1)/2$ such failure combinations).

The failure of the first subsystem will occur with a probability of $(1-a)$ as expected. **However, once one subsystem has failed, the system then behaves as if it comprises $n-1$ remaining subsystems with decreased availability.**

Subsystem availability is reduced by a subtractive term. Note that as system recovery time, R, goes to zero, a' approaches a. Thus, minimizing recovery time is paramount for improving system availability in the face of failover faults.

The above interpretation can be applied to all of the previous models. The expanded results for all models are given in Table 11-2.

Repair	Failover Fault	
	None	With Subsystem Repair
Parallel	a	$a - \dfrac{2p}{n-1}$
Sequential	a	$a - \dfrac{p}{n-1}$
Simultaneous	a	$a - \dfrac{p}{n-1}$

Repair	With System Recovery
Parallel	$a - \left(\dfrac{R}{r/2 + R}\right)\dfrac{p}{n-1}$
Sequential	$a - \left(\dfrac{R}{r + R}\right)\dfrac{p}{n-1}$
Simultaneous	$a - \left(\dfrac{R}{r + R}\right)\dfrac{p}{n-1}$

a'
**Effective Subsystem Availability
Following One Subsystem Failure
Table 11-2**

To summarize the notation in this table,

    N is the number of subsystems in the system.
    A is the subsystem availability.
    R is the subsystem average repair time.
    R is the system average recovery time.
    P is the probability of a failover fault.

System Availability Degradation

The previous results offer another interesting insight. Consider the case represented in Table 11-2 for a system that requires only subsystem repair and that uses sequential repair:

$$F \approx n(n-1)(1-a)(1-a') = n(n-1)(1-a)\left[1-\left(a-\frac{p}{n-1}\right)\right] \quad (11\text{-}11)$$

Taking the two subsystem case ($n=2$),

$$F \approx 2(1-a)[1-(a-p)] = 2(1-a)[(1-a)+p] \quad (11\text{-}12)$$

$(1-a)$ is the probability of failure of a subsystem, and p is the probability of the failure of a failover. If p is very much larger than $(1-a)$, then

$$F \approx 2(1-a)p \quad (11\text{-}13)$$

Thus, the system degrades to the equivalent of one subsystem with a failure probability of $(1-a)$ and one with a much higher failure probability of p as failover faults become prominent. This states the obvious: the probability of the first failure is that of a subsystem failure, and the probability of the second failure is that of a failover fault.

For the suggested values of a = .999 and $(1-p)$ = .99, this will reduce system availability by one 9 - an order of magnitude increase in the failure probability. This shows the importance of proper failover, both from a software viewpoint and from a human viewpoint.

Splitting Systems

Consider a system with n subsystems that does not suffer failover faults (sequential repair is assumed as an example, but the conclusions do not depend on this). Its failure probability F is

$$F \approx n(n-1)(1-a)^2$$

If we were to split this system into k independent partitions of n/k partitions each, then each partition will have a failure probability of

$$\frac{n}{k}\left(\frac{n}{k}-1\right)(1-a)^2$$

However, since there are now k partitions, the probability of failure, F_p, for the entire partitioned system is k times this value or

$$F_p \approx \frac{n}{k}(n-k)(1-a)^2$$

Comparing these two values as a ratio, we have

$$\frac{F}{F_p} \approx k\frac{n-1}{n-k} > k \qquad (11\text{-}14)$$

Thus, partitioning the system into k partitions increases its reliability by a factor of at least k.

However, this advantage is eroded if there is a chance for a failover fault. Using systems that require only subsystem repair with no recovery (with sequential repair) as an example, Equation (11-5) tells us that

$$F \approx n(n-1)(1-a)^2 + pn(1-a)$$

If we partition this system into k partitions, the failure probability of the partitioned system is

$$F_p \approx k\left[\frac{n}{k}(\frac{n}{k}-1)(1-a)^2 + p\frac{n}{k}(1-a)\right] = \frac{n}{k}(n-k)(1-a)^2 + pn(1-a)$$

Comparing these two failure probabilities, we have

Subsystem repair, sequential/simultaneous repair:

$$\frac{F}{F_p} \approx k \frac{(n-1) + \frac{p}{1-a}}{(n-k) + \frac{kp}{1-a}} \qquad (11\text{-}15a)$$

If there are no failover faults ($p = 0$), then this reduces to the reliability enhancement of Equation (11-14). However, as the probability p of a failover fault approaches the probability of a subsystem failure ($1-a$), the reliability advantage of system splitting decreases and ultimately disappears for $p \gg (1-a)$.

Thus, system splitting is only advantageous from a reliability viewpoint if p is small. Specifically, kp should be much less than ($1-a$) in order for system splitting to achieve its full potential.

If parallel repair were used instead in the above example, then

Subsystem repair, parallel repair:

$$\frac{F}{F_p} \approx k \frac{(n-1) + \frac{2p}{1-a}}{(n-k) + \frac{2kp}{1-a}} \qquad (11\text{-}15b)$$

For failover faults with system recovery,

System recovery, sequential/simultaneous repair:

$$\frac{F}{F_p} \approx k \frac{(n-1) + \frac{R}{r+R}\frac{p}{1-a}}{(n-k) + \frac{R}{r+R}\frac{kp}{1-a}} \qquad (11\text{-}15c)$$

System recovery, parallel repair:

Dr. Bill Highleyman, Paul J. Holenstein, and Dr. Bruce Holenstein

$$\frac{F}{F_p} \approx k \frac{(n-1) + \dfrac{R}{r/2 + R} \dfrac{2p}{1-a}}{(n-k) + \dfrac{R}{r/2 + R} \dfrac{2kp}{1-a}} \qquad (11\text{-}15d)$$

Interestingly, for those cases that require system recovery (Equations (11-15c) and (15d)), the impact of failover faults in a split system is not only a function of the failover fault probability p but is also a function of system recovery time, R. If the system recovery time is zero, then failover faults have no adverse affect on the reliability increase obtainable by splitting a system. In effect, if $R = 0$, a failover fault does not cause any system down time because it is immediately restored to service. Again, minimizing system recovery time is paramount to improving system availability.

Note that it is inappropriate to study the above relations (15c) and (15d) and conclude that the impact of failover faults can be minimized by increasing subsystem repair time r. This says simply that the impact of failover faults is minimized by making the subsystems less reliable – a true statement but one we will be unlikely to follow.

An example will illustrate the impact of failover faults on system splitting. Let us assume that we want to split a 16-processor system into four 4-processor nodes. If failover faults were not a consideration, this would give us a five-fold increase in reliability (4 x 15/12). However, let us consider a more realistic situation in which

subsystem failure probability (1-a) is	.001
failover fault probability (p) is	.01
subsystem repair time (r) is	24 hours
system recovery time (R) is	4 hours
parallel repair (Equation 11-15d) is used	

Then the reliability advantage of the split system compared to the single un-split system is reduced from 5 to 3.18.

The Importance of System Recovery Time

The above considerations have pointed out the importance of system recovery time to system availability. This can be further emphasized through the following observation.

A single CPU halt can be caused either by hardware or by software. It appears from current experience that, more or less,

- 20% of all CPU halts are caused by hardware.
- 80% of all CPU halts are caused by software or human error.

It takes two CPU halts to cause a system outage. One can then conclude that only 4% (.2 x .2) of all system outages are caused by dual hardware faults. 96% of all system outages are caused either by a combination of a hardware fault and a software fault or by two software faults. These cases require no hardware repair – only a system recovery.

Therefore, a repair is not needed for the vast majority of system outages. Given the above scenario for today's environment, system recovery time is the predominant factor in system availability. Hardware repair is a relatively minor factor.

Markov Modeling

Much more accurate expressions for these cases may be obtained from Markov models. They are evaluated for each case in Appendix 3.

A Markov model defines the various states in which a system can exist and defines the transitions for each state and the probability of those transitions. For instance, Figure 11-1 shows a State B that can be entered from State A and can exit to State C. While in State A, the system will transition to State B at a rate of r_{ab}. While in state B, it will transition to State C at a rate of r_{bc}.

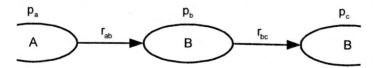

State Transitions
Figure 11-1

If the probability of the system being in State A is p_a and in State B is p_b, then the rate of the system transitioning to State B is $p_a r_{ab}$ and from State B is $p_b r_{bc}$. The difference between these rates is the rate of change of the probability that the system will be in State B:

$$\frac{dp_b}{dt} = p_a r_{ab} - p_b r_{bc}$$

In the steady state, $\frac{dp_b}{dt}=0$; and the incoming transition rate must equal the exiting transition rate:

$$p_a r_{ab} = p_b r_{bc}$$

State transition equations such as this can be written for each state in the system. This yields n equations with n variables (the state probabilities). However, these equations are not independent; summing any n-1 equations will result in the remaining equation. The nth independent equation is the summation of all probabilities, which must equal 1:

$$\sum_{i=0}^{n} p_i = 1$$

Let us evaluate a simple model as an example. The case we will model is that of a system comprising two subsystems. Parallel repair is assumed. The system can survive a single subsystem failure. The Markov model for this case is shown in Figure 11-2.

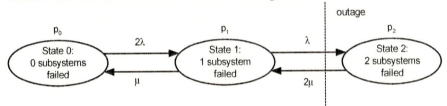

Dual Subsystem Model
Figure 11-2

There are three states:

(1) The system is in State 0 if no subsystems have failed. It will be in this state with probability p_o.

(2) The system is in State 1 if one subsystem has failed. It will be in this state with probability p_1.

(3) The system is in State 2 if two subsystems have failed. It will be in this state with probability p_2. This state represents a system outage.

Dr. Bill Highleyman, Paul J. Holenstein, and Dr. Bruce Holenstein

Let λ be the rate of failure of a subsystem and μ be its rate or repair. Note that λ and μ are the inverses of the subsystem mtbf and mtr, respectively:

$$\lambda = \frac{1}{\text{mtbf}}$$
$$\mu = \frac{1}{\text{mtr}}$$

Therefore, in State 0, the failure rate is 2λ since there are two subsystems that can fail. In State 1, the failure rate is λ and the repair rate is u since there is one subsystem that is good and that can fail, and one subsystem that is failed and can be repaired. In State 2, the repair rate is $2u$ since there are two subsystems that can be repaired.

The state relations are:

State 0: $\quad \mu p_1 = 2\lambda p_0$

State 1: $\quad 2\lambda p_0 + 2\mu p_2 = (\mu + \lambda) p_1$

State 2: $\quad \lambda p_1 = 2\mu p_2$

Using

$$p_0 + p_1 + p_2 = 1$$

the resulting state probabilities are:

$$p_0 = \frac{\mu^2}{\lambda^2 + 2\mu\lambda + \mu^2}$$

$$p_1 = \frac{2\mu\lambda}{\lambda^2 + 2\mu\lambda + \mu^2}$$

$$p_2 = \frac{\lambda^2}{\lambda^2 + 2\mu\lambda + \mu^2}$$

Note that

$$\frac{\mu}{\lambda} = \frac{mtbf}{mtr}$$

$$a = \frac{mtbf}{mtbf + mtr} = \frac{1}{1 + \frac{\lambda}{\mu}}$$

Thus,

$$\frac{\mu}{\lambda} = \frac{a}{1-a}$$

Using this relation, the above state probabilities can be expressed in terms of subsystem availability a:

$$p_0 = a^2$$
$$p_1 = 2a(1-a)$$
$$p_2 = (1-a)^2$$

The various cases discussed in this chapter are formally evaluated using Markov models in the following Appendix 3. Following each model evaluation, an approximation of the system failure probability is given. This approximation is based on the assumptions that

$$a^i \approx 1$$
$$(1-a) \ll 1$$
$$p \ll 1$$

The approximations agree with the intuitive models presented in the body of this chapter.

Appendices

Appendix 1 - Availability Relationships

A variety of relationships associated with system availability have been derived in this book. They are summarized here for convenient access, with reference to the appropriate point of derivation given by the equation numbers. The parameters used by these relationships are as follows:

A	system availability (proportion of time system is operational)
a	subsystem availability (proportion of time subsystem is operational)
D	number of lockable items in the database
d	number of databases in the application network
e	comparative synchronous replication efficiency (coordinated commit efficiency / dual write efficiency)
F	probability of system failure (proportion of time system is down)
f	number of failure modes for a system
k	number of nodes in a system
L	replication latency time
L_{cc}	application latency for coordinated commits
L_{dw}	application latency for dual writes
MTBF	mean time before system failure
mtbf	mean time before subsystem failure
MTR	mean time to repair/recover/restore system
mtr	mean time to repair/recover/restore subsystem
n	number of subsystems in a system
n_u	average number of updates in a transaction
n_{u1}	number of transaction operations that require one communication channel round trip

n_{u2}	number of transaction operations that require two communication channel round trips
p	round trip communication channel time ($2t_c$) expressed as a proportion of replication delay time t_p
R	mean time to recover a system.
r	mean time to repair a subsystem.
R_w	transaction wait rate (synchronous replication).
R_d	transaction deadlock rate (synchronous replication).
R_c	transaction collision rate (asynchronous replication).
s	number of spares
t	average time required to complete an action
t_c	communication channel propagation time
t_o	average database operation time for the particular mix of operations generated by a transaction
t_p	processing delay through the replicator exclusive of the communication channel propagation time

Availability

$$A = \frac{MTBF}{MTBF + MTR} \tag{1-1a}$$

$$A = \frac{1}{1 + \frac{MTR}{MTBF}} \approx 1 - \frac{MTR}{MTBF} \tag{1-1b}$$

$$F = 1 - A \approx \frac{MTR}{MTBF} \tag{1-2}$$

$$MTBF \approx MTR/(1-A) \tag{1-3}$$

$$MTR \approx MTBF(1-A) \qquad (1\text{-}4)$$

$$A = a_1 a_2 \qquad (1\text{-}5)$$
(if both subsystems 1 and 2 must be operational)

$$A = 1 - (1-a)^2 \qquad (1\text{-}6)$$
(if either subsystem 1 or 2 must be operational)

$$A = 1 - F \approx 1 - f(1-a)^{s+1} \qquad (1\text{-}7)$$

$$f = \binom{n}{s+1} = \frac{n!}{(s+1)!(n-s-1)!} \qquad (2\text{-}2a)$$

$$f = \frac{n(n-1)}{2} \quad \text{(for } n = 2\text{)} \qquad (2\text{-}2b)$$

$$A = 1 - F \approx 1 - f\left(\frac{mtr}{mtbf}\right)^{s+1} \qquad (1\text{-}9)$$

$$MTR = \frac{mtr}{(s+1)} \qquad (1\text{-}10)$$

$$MTBF \approx \frac{mtbf}{f(s+1)}\left(\frac{mtbf}{mtr}\right)^s \qquad (1\text{-}11)$$

System Splitting

$$\text{reliability ratio} = k\frac{n-1}{n-k} > k \qquad (2\text{-}11)$$

Synchronous Replication

$$L_{dw} = (4n_u + 4)t_c \tag{4-1}$$

$$L_{cc} = t_p + 2t_c \tag{4-2}$$

$$e = \frac{2(n_u + 1)}{1 + 1/p} \tag{4-4}$$

$$p = 2t_c / t_p \tag{4-5}$$

$$p = 1/(2n_u + 1) \quad (\text{for } e = 1) \tag{4-6}$$

$$n_u = n_u' = n_{u2} + n_{u1}/2 \tag{4-7}$$
(if some operations require only one round trip)

$$L_{dw} = (4n_u' + 4)t_c + n_u t_o \tag{4-1a}$$
(for dual write serial updates)

$$e = \frac{2(n_u' + 1) + \dfrac{n_u t_o}{2 t_c}}{1 + 1/p} \tag{4-4a}$$
(for dual write serial updates)

$$L_{dw} = (4n_u' + 4)t_c + (d-1)n_u t_o \tag{4-1b}$$
(for plural writes)

$$e = \frac{2(n_u' + 1) + (d-1)\dfrac{n_u t_o}{2 t_c}}{1 + 1/p} \tag{4-4b}$$
(for plural writes)

Failover Faults

For the failover fault relationships, the following term is redefined:

p probability of a failover fault

$$F \approx \frac{r/2 + R}{r/2} \frac{n(n-1)}{2}(1-a)^2 + \frac{R}{r}pn(1-a) \qquad \begin{array}{c}(5\text{-}4c)\\(11\text{-}6)\end{array}$$
(parallel repair)

$$F \approx \frac{r+R}{r} n(n-1)(1-a)^2 + \frac{R}{r}pn(1-a) \qquad (11\text{-}7)$$
(sequential, simultaneous repair)

$$F \approx \frac{r/2 + R}{r/2} \frac{n(n-1)}{2}(1-a)(1-a') \qquad \begin{array}{c}(5\text{-}5)\\(11\text{-}12)\end{array}$$
(parallel repair)

$$a' = a - \frac{R}{r/2 + R} \frac{p}{n-1} \qquad \begin{array}{c}(5\text{-}6)\\(11\text{-}14)\end{array}$$

$$\text{reliability ratio} = k\frac{(n-1)+x}{(n-k)+kx} \qquad (5\text{-}8)$$
(for split system)

$$x = \frac{R}{r/2 + R} \frac{2p}{1-a} \qquad \begin{array}{c}(5\text{-}9)\\(11\text{-}15d)\end{array}$$

Dr. Bill Highleyman, Paul J. Holenstein, and Dr. Bruce Holenstein

Data Conflict Rates

For the data conflict relationships, the following terms are redefined:

- a for asynchronous replication, the number of modification actions that may cause a collision in an average transaction (such as inserts, updates, deletes); for synchronous replication, the total number of modification actions in a transaction.
- a' the total number of modification actions in an average transaction (whether or not they can cause a collision under asynchronous replication). For synchronous replication, $a = a'$.
- r application transaction rate arriving at each node

$$R_w = \frac{d(kra)^2 at}{2D} \tag{9-26}$$
(synchronous replication)

$$R_d = \frac{d(kra)^2 a^3 t}{4D^2} \tag{9-27}$$
(synchronous replication - mutual waits)

$$R_d = \frac{d-1}{d}\frac{(kra)^2}{D}S \tag{9-28}$$
(synchronous replication - lock latency)

$$R_c = \frac{(d-1)}{d}\frac{(kra)^2}{D}S \tag{9-29}$$
(asynchronous replication)

$$S = \frac{d(a't+L)}{2} \qquad (9\text{-}19a)$$

(transactions sent serially after commit)

$$S = a't + L \qquad (9\text{-}20a)$$

(transactions broadcast after commit)

$$S = \frac{d(t+L)}{2} \qquad (9\text{-}21)$$

(modifications sent serially after application)

$$S = t + L \qquad (9\text{-}22)$$

(modifications broadcast after application)

$$S = L \qquad (9\text{-}23)$$

(modifications broadcast upon receipt)

Appendix 2 - Availability Approximation Analysis

The availability relationships developed earlier are noted to be approximations. But how good are these approximations?

Consider a system of n identical elements arranged such that s of these elements are spares. That is, $(n-s)$ elements must be operational in order for the system to be operational. The probability that an element will be operational is denoted by a:

a = probability that a system element is operational.

At any point in time, the system may be in one of many states. All n elements may be operational, $n-1$ elements may be operational with one failed element, and so on to the state where all elements have failed.

Assuming that element failures are independent of each other, then the probability that n elements will be operational is a^n, the probability that a specific set of $n-1$ elements will be operational is $a^{n-1}(1-a)$ (that is, $n-1$ elements are operational; and one has failed) and so on. Let f_i be the number of ways in which i different elements can fail (that is, the number of different system states leading to $n-i$ operational elements and i failed elements):

i number of failed elements
f_i number of ways in which exactly i elements can fail.

Then the probability that the system state will be that of i failed elements is:

$$f_i a^{n-i}(1-a)^i$$

f_i is the number of ways that i elements can be chosen from n elements:

$$f_i = \binom{n}{i} = \frac{n!}{i!(n-i)!}$$

Since the range of i from 0 to n represents the universe of system states, then it follows that

$$\sum_{i=0}^{n} \binom{n}{i} a^{n-i}(1-a)^i = 1$$

Since there are s spares in the system, only those states for which $i > s$ can represent system failures. Furthermore, for any given number i of element failures, not all f_i combinations may result in a system failure. Perhaps the system may survive some combinations of i failures even though this exceeds the number of spares. Let f_i' be the actual number of combinations of i failures that will lead to a system failure:

f_i' = number of combinations of i failures that will cause a system failure.

Then the probability of system failure, F, is

$$F = \sum_{i=s+1}^{n} f_i' a^{n-i}(1-a)^i \qquad (A2\text{-}1)$$

If a is very close to 1 so that $(1-a)$ is very small, then only the first term of Equation (A2-1) is significant (this depends on f_i' not being a strong function of i, which is usually the case). Equation (A2-1) can then be approximated by

$$F \approx f_{s+1}' a^{n-s-1}(1-a)^{s+1} \qquad (A2\text{-}2)$$

Furthermore, since a is very close to 1 (and if n-s-1 is not terribly large), then

$$a^{n-s-1} \approx 1$$

Defining f to be f'_{s+1}, then Equation (A2-2) can be further approximated by

$$F \approx f(1-a)^{s+1} \qquad (A2\text{-}3)$$

and system availability A is approximately

$$A \approx 1 - f(1-a)^{s+1} \qquad (A2\text{-}4)$$

where

- a is the availability of a system element
- s is the number of spare elements
- f is the number of ways in which $s+1$ elements can fail in such a way as to cause a system failure
- F is the probability of failure of the system
- A is the availability of the system

Equation (A2-4) is the same as Equation (1-1) derived heuristically in Chapter 1.

A feel for the degree of approximation afforded by Equation (A2-4) is shown in the following table for $a=.995$, n ranging from 2 through 16, $f = f_i$, and s ranging from 0 through $n-1$. The table shows that the maximum approximation error of Equation (A2-4) as compared to the exact representation of Equation (A2-1) does not exceed 5% over this range of parameters.

The value of this approximation lies not so much in its calculation ease (especially in today's world of spreadsheets) as it does in the insight it provides about the roles that failure modes, sparing, and element reliability play in system availability.

n	s	f	F (Eq. A-1)	A (1-F)	approx F (Eq. A-4)	(approx F % error - high if > 0)
2	0	2	0.009975	0.990025	0.01	0.25
2	1	1	0.000025	0.999975	0.000025	0.00
4	0	4	0.019850499	0.980150	0.02	0.75
4	1	6	0.000149002	0.999851	0.00015	0.67
4	2	4	4.98125E-07	1	5E-07	0.38
4	3	1	6.25E-10	1	6.25E-10	0.00
8	0	8	0.039306956	0.960693	0.04	1.76
8	1	28	0.000686131	0.999314	0.0007	2.02
8	2	56	6.8698E-06	0.999993	7E-06	1.90
8	3	70	4.30544E-08	1	4.375E-08	1.62
8	4	56	1.72822E-10	1	1.75E-10	1.26
8	5	28	4.33758E-13	1	4.375E-13	0.86
8	6	8	6.22266E-16	1	6.25E-16	0.44
8	7	1	3.90625E-19	1	3.906E-19	0.00
16	0	16	0.077068876	0.922931	0.08	3.80
16	1	120	0.002863359	0.997137	0.003	4.77
16	2	560	6.66682E-05	0.999933	7E-05	5.00
16	3	1820	1.08413E-06	0.999999	1.138E-06	4.92
16	4	4368	1.30376E-08	1	1.365E-08	4.70
16	5	8008	1.19867E-10	1	1.251E-10	4.39
16	6	11440	8.59178E-13	1	8.938E-13	4.02
16	7	12870	4.85138E-15	1	5.027E-15	3.63
16	8	11440	2.16494E-17	1	2.234E-17	3.21
16	9	8008	7.60946E-20	1	7.82E-20	2.77
16	10	4368	2.08438E-22	1	2.133E-22	2.32
16	11	1820	4.3619E-25	1	4.443E-25	1.87
16	12	560	6.74117E-28	1	6.836E-28	1.41
16	13	120	7.25602E-31	1	7.324E-31	0.94
16	14	16	4.85992E-34	1	4.883E-34	0.47
16	15	1	1.52588E-37	1	1.526E-37	0.00

Appendix 3 - Failover Fault Models

General

In this Appendix, the exact state solutions for a single-spared redundant system comprising n identical subsystems are derived for the following cases:

No failover fault:
    Parallel repair
    Sequential repair
    Simultaneous repair
Failover fault requiring subsystem repair:
    Parallel repair
    Sequential repair
    Simultaneous repair
Failover fault requiring system recovery:
    Parallel repair
    Sequential repair
    Simultaneous repair

The repair strategies are further explained in Chapter 11.

The following notation is used:

λ	subsystem failure rate
μ	subsystem repair rate
α	$= \mu/\lambda$
γ	system recovery rate
a	subsystem availability
n	number of subsystems in the system
i	number of failed subsystems
p_i	probability of state i

p	probability of failover fault
p_f	probability of failover fault state
p_r	probability of recovery state
F	probability of system outage
mtr	subsystem mean time to repair = $1/\mu$
mtbf	subsystem mean time before failure = $1/\lambda$
r	subsystem mean time to repair = mtr = $1/\mu$
R	system mean time to recover = $1/\gamma$

$\binom{n}{i} = \dfrac{n!}{i!(n-i)!}$, the number of ways that of i out of n subsystems can fail

The following relationships are used:

$$\alpha = \frac{\mu}{\lambda} = \frac{mtbf}{mtr} = \frac{\frac{mtbf}{mtbf+mtr}}{1-\frac{mtbf}{mtbf+mtr}} = \frac{a}{1-a} \qquad (A3\text{-}1)$$

Using the Binomial Theorem,

$$\sum_{i=0}^{n}\binom{n}{i}a^{n-i}(1-a)^{i} = [a+(1-a)]^{n} = 1 \qquad (A3\text{-}2)$$

With this result, we can further state that

$$\sum_{i=0}^{n}\frac{n!}{(n-i)!}a^{n-i}(1-a)^{i} = \sum_{i=0}^{n}\binom{n}{i}a^{n-i}(1-a)^{i} + \sum_{i=0}^{n}(i!-1)\binom{n}{i}a^{n-i}(1-a)^{i}$$

or

$$\sum_{i=0}^{n}\frac{n!}{(n-i)!}a^{n-i}(1-a)^{i} = 1 + \sum_{i=2}^{n}(i!-1)\binom{n}{i}a^{n-i}(1-a)^{i} \qquad (A3\text{-}3)$$

where the summation lower bound was changed to 2 since $(i!-1) = 0$ for $i=0$ and 1.

The following approximations are made in order to reduce the complete models developed in the following sections to the intuitive approximate models developed in the body of this book:

$a^i \approx 1$
$(1-a) \ll 1$
$p \ll 1$
$\lambda \ll \mu$
$\alpha = \dfrac{\mu}{\lambda} = \dfrac{a}{1-a} \gg 1$

No Failover Fault

Parallel Repair

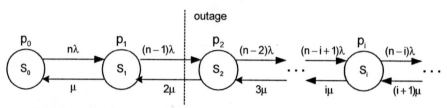

Description:

In State 0, n subsystems are operational and fail at a rate of λ each. A subsystem failure causes an entry into State 1. From State 1, the failed subsystem may be repaired at a rate of μ, and State 0 is reentered. Alternatively, another operational subsystem may fail. Since there an n-1 operational subsystems in State 1, subsystem failure rate is $(n-1)\lambda$. When a subsystem fails in State 1, State 2 is entered. In this state, there are two failed subsystems, each being repaired at a rate of μ. Thus, repairs occur at a rate of 2μ in State 2. When a subsystem is repaired, State 1 is reentered. Failures in State 2 occur at a rate of $(n-2)\lambda$. When a failure occurs in State 2, State 3 is entered. This pattern continues for States 3 through n.

State Transitions:

State	State Transition Equation
0	$\mu p_1 = n\lambda p_0$
1	$n\lambda p_0 + 2\mu p_2 = [(n-1)\lambda + \mu]p_1$
2	$(n-1)\lambda p_1 + 3\mu p_3 = [(n-2)\lambda + 2\mu]p_2$
3 - n	$(n-i+1)\lambda p_{i-1} + (i+1)\mu p_{i+1} = [(n-i)\lambda + i\mu]p_i$

State Probabilities (in terms p_0):

State	State Probability
0	p_0
1	$n\alpha^{-1}p_0$
2	$\dfrac{n(n-1)}{2}\alpha^{-2}p_0$
3 - n	$\dfrac{n!}{i!(n-i)!}\alpha^{-i}p_0 = \binom{n}{i}\alpha^{-i}p_0$
where	$\alpha = \dfrac{\mu}{\lambda}$

State Probabilities (in terms of failure/repair rates):

State	State Probability
0	$\dfrac{\alpha^n}{D}$
1	$\dfrac{n\alpha^{n-1}}{D}$
2	$\dfrac{\frac{n(n-1)}{2}\alpha^{n-2}}{D}$
3 - n	$\dfrac{\frac{n!}{i!(n-i)!}\alpha^{n-i}}{D}$
where	$D = \sum\limits_{i=0}^{n}\binom{n}{i}\alpha^i$

State Probabilities (in terms of subsystem availability a):

State	State Probability
0	$\dfrac{a^n}{D'}$
1	$\dfrac{na^{n-1}(1-a)}{D'}$

2	$\dfrac{\dfrac{n(n-1)}{2}a^{n-2}(1-a)^2}{D'}$
3 - n	$\dfrac{\binom{n}{i}a^{n-i}(1-a)^i}{D'}$
where	$D' = (1-a)^n D = \sum_{i=0}^{n}\binom{n}{i}a^{n-i}(1-a)^i = 1$

using α=a/(1-a) and relation (A3-2)

System Failure Probability:

$$F \approx p_2 = \frac{n(n-1)}{2}a^{n-2}(1-a)^2 \approx \frac{n(n-1)}{2}(1-a)^2$$

Breaking the Availability Barrier

Sequential Repair

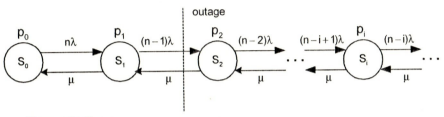

Description:

In State 0, n subsystems are operational and fail at a rate of λ each. A subsystem failure causes an entry into State 1. From State 1, the failed subsystem may be repaired at a rate of μ; and State 0 is reentered. Alternatively, another operational subsystem may fail. Since there an n-1 operational subsystems in State 1, the subsystem failure rate is $(n-1)\lambda$. When a subsystem fails in State 1, State 2 is entered. In this state, there are two failed subsystems being repaired one at a time at a rate of μ. Thus, repairs occur at a rate of μ in State 2. When a subsystem is repaired, State 1 is reentered. Failures in State 2 occur at a rate of $(n-2)\lambda$. When a failure occurs in State 2, State 3 is entered. This pattern continues for States 3 through n.

State Transitions:

State	State Transition Equation
0	$\mu p_1 = n\lambda p_0$
1	$n\lambda p_0 + \mu p_2 = [(n-1)\lambda + \mu]p_1$
2	$(n-1)\lambda p_1 + \mu p_3 = [(n-2)\lambda + \mu]p_2$
3 - n	$(n-i+1)\lambda p_{i-1} + \mu p_{i+1} = [(n-i)\lambda + \mu]p_i$

State Probabilities (in terms p_0):

State	State Probability
0	p_0
1	$n\alpha^{-1} p_0$
2	$n(n-1)\alpha^{-2} p_0$

311

Dr. Bill Highleyman, Paul J. Holenstein, and Dr. Bruce Holenstein

State	
3 - n	$\dfrac{n!}{(n-i)!}\alpha^{-i} p_0$
where	$\alpha = \dfrac{\mu}{\lambda}$

State Probabilities (in terms of failure/repair rates):

State	State Probability
0	$\dfrac{\alpha^n}{D}$
1	$\dfrac{n\alpha^{n-1}}{D}$
2	$\dfrac{n(n-1)\alpha^{n-2}}{D}$
3 - n	$\dfrac{\dfrac{n!}{(n-i)!}\alpha^{n-i}}{D}$
where	$D = \sum\limits_{i=0}^{n} \dfrac{n!}{(n-i)!}\alpha^{n-i}$

State Probabilities (in terms of subsystem availability a):

State	State Probability
0	$\dfrac{a^n}{D'}$
1	$\dfrac{na^{n-1}(1-a)}{D'}$
2	$\dfrac{n(n-1)a^{n-2}(1-a)^2}{D'}$
3 - n	$\dfrac{\dfrac{n!}{(n-i)!}a^{n-i}(1-a)^i}{D'}$

where $D' = (1-a)^n D = \sum\limits_{i=0}^{n} \dfrac{n!}{(n-i)!} a^{n-i}(1-a)^i = 1 + \sum\limits_{2}^{n}(i!-1)\binom{n}{i} a^{n-i}(1-a)^i \approx 1$

using $\alpha = a/(1-a)$, relation (A3-3), and $(1-a) \ll 1$

System Failure Probability:

$$F \approx p_2 = \frac{n(n-1)a^{n-2}(1-a)^2}{D'} \approx n(n-1)(1-a)^2$$

Simultaneous Repair

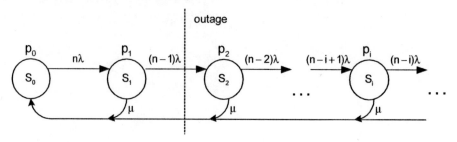

Description:

In State 0, n operational subsystems fail at a rate of λ each. A subsystem failure causes an entry into State 1. From State 1, the failed subsystem may be repaired at a rate of μ; and State 0 is reentered. Alternatively, another operational subsystem may fail. Since there an n-1 operational subsystems in State 1, the subsystem failure rate is $(n-1)\lambda$. When a subsystem fails in State 1, State 2 is entered. In this state, there are two failed subsystems, which are repaired simultaneously. Thus, repairs occur at a rate of μ in State 2; and State 0 is reentered since now all subsystems are operational. This pattern continues for States 3 through n.

State Transitions:

State	State Transition Equation
0	$\mu \sum_{i=1}^{n} p_i = n\lambda p_0$
1	$n\lambda p_0 = [(n-1)\lambda + \mu]p_1$
2	$(n-1)\lambda p_1 = [(n-2)\lambda + \mu]p_2$
3 - n	$(n-i+1)\lambda p_{i-1} = [(n-i)\lambda + \mu]p_i$

State Probabilities (in terms p_0):

State	State Probability
0	p_0
1	$\dfrac{n}{(n-1)+\alpha} p_0$

Breaking the Availability Barrier

State	
2	$\dfrac{n(n-1)}{[(n-1)+\alpha][(n-2)+\alpha]} p_0$
3 - n	$\dfrac{\dfrac{n!}{(n-i)!}}{\prod_{j=1}^{i}[(n-j)+\alpha]} p_0$
where	$\alpha = \dfrac{\mu}{\lambda}$

State Probabilities (in terms of failure/repair rates):

State	State Probability
0	α^n / D
1	$\dfrac{n\alpha^n}{(n-1)+\alpha} / D$
2	$\dfrac{n(n-1)\alpha^n}{[(n-1)+\alpha][(n-2)+\alpha]} / D$
3 - n	$\dfrac{\dfrac{n!}{(n-i)!} \alpha^n}{\prod_{j=1}^{i}[(n-j)+\alpha]} / D$
where	$D = \alpha^n + \displaystyle\sum_{i=1}^{n} \dfrac{n!}{(n-i)!} \dfrac{\alpha^n}{\prod_{j=1}^{i}[(n-j)+\alpha]} \approx \displaystyle\sum_{0}^{n} \dfrac{n!}{(n-i)!} \alpha^{n-i}$
	for $\alpha \gg 1$

State Probabilities (in terms of subsystem availability a):

State	State Probability
0	$\dfrac{a^n}{D'}$
1	$\dfrac{na^{n-1}(1-a)}{D'}$
2	$\dfrac{n(n-1)a^{n-2}(1-a)^2}{D'}$

3 - n $\dfrac{\dfrac{n!}{(n-i)!}a^{n-i}(1-a)^i}{D'}$

where
(1) $D' = (1-a)^n D \approx \sum_{i=0}^{n} \dfrac{n!}{(n-i)!} a^{n-i}(1-a)^i = 1 + \sum_{2}^{n}(i!-1)\binom{n}{i}a^{n-i}$

using α = a/(1-a), relation (A3-3), and (1-a) << 1

System Failure Probability:

$F \approx p_2 \approx \dfrac{n(n-1)a^{n-2}(1-a)^2}{D'} \approx n(n-1)(1-a)^2$

Failover Fault Requiring Subsystem Repair

Parallel Repair

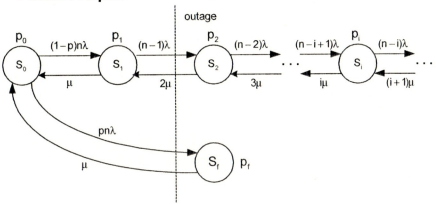

Description:

In State 0, n operational subsystems fail at a rate of λ each. (1-p) of the time, a subsystem failure causes an entry into State 1. From State 1, the failed subsystem may be repaired at a rate of μ; and State 0 is reentered. Alternatively, another operational subsystem may fail. Since there an n-1 operational subsystems in State 1, the subsystem failure rate is $(n-1)\lambda$. When a subsystem fails in State 1, State 2 is entered. In this state, there are two failed subsystems, each being repaired at a rate of μ. Thus, repairs occur at a rate of 2μ in State 2. When a subsystem is repaired, State 1 is reentered. This sequence continues for States 3 through n.

Alternatively, a subsystem failure in State 0 will experience a failover fault with a probability of p. In this case, the system enters State f where the failed subsystem is repaired at a rate of μ; and State 0 is then reentered.

State Transitions:

State	State Transition Equation
0	$\mu p_1 + \mu p_f = n\lambda p_0$
1	$(1-p)n\lambda p_0 + 2\mu p_2 = [(n-1)\lambda + \mu]p_1$
2	$(n-1)\lambda p_1 + 3\mu p_3 = [(n-2)\lambda + 2\mu]p_2$
3 - n	$(n-i+1)p_{i-1} + (i+1)\mu p_{i+1} = [(n-i)\lambda + i\mu]p_i$
f	$pn\lambda p_0 = \mu p_f$

State Probabilities (in terms p_0):

State	State Probability
0	p_0
1	$(1-p)n\alpha^{-1}p_0$
2	$(1-p)\dfrac{n(n-1)}{2}\alpha^{-2}p_0$
3 - n	$(1-p)\dfrac{n!}{i!(n-i)!}\alpha^{-i}p_0$
f	$pn\alpha^{-1}p_0$
where	$\alpha = \dfrac{\mu}{\lambda}$

Failure State Probability (in terms of failure/repair rates):

State	State Probability
0	$\dfrac{\alpha^n}{D}$
1	$(1-p)n\dfrac{\alpha^{n-1}}{D}$
2	$(1-p)\dfrac{n(n-1)}{2}\dfrac{\alpha^{n-2}}{D}$
3 - n	$(1-p)\dfrac{n!}{i!(n-i)!}\dfrac{\alpha^{n-i}}{D}$
f	$p_f = pn\dfrac{\alpha^{n-1}}{D}$

where $$D = pn\alpha^{n-1} + p\alpha^n + (1-p)\sum_{0}^{n}\binom{n}{i}\alpha^{n-i}$$

Failure State Probability (in terms of subsystem availability a):

State	State Probability
0	$\dfrac{a^n}{D'}$
1	$(1-p)\dfrac{na^{n-1}(1-a)}{D'}$
2	$(1-p)\dfrac{\frac{n(n-1)}{2}a^{n-2}(1-a)^2}{D'}$
3 - n	$(1-p)\dfrac{\binom{n}{i}a^{n-i}(1-a)^i}{D'}$
f	$p\dfrac{na^{n-1}(1-a)}{D'}$

$$D' = (1-a)^n D$$

where
$$= pna^{n-1}(1-a) + pa^n + (1-p)\sum_{i=0}^{n}\binom{n}{i}a^{n-i}(1-a)^i$$

$$\approx 1 - p(1-a^n) \approx 1$$

using $\alpha = a/(1-a)$, relation (A3-2), $(1-a) \ll 1$, and $p \ll 1$

System Failure Probability:

$$F \approx p_2 + p_f = \dfrac{(1-p)\dfrac{n(n-1)}{2}a^{n-2}(1-a)^2 + pna^{n-1}(1-a)}{D'}$$

$$\approx \dfrac{n(n-1)}{2}(1-a)^2 + pn(1-a)$$

Sequential Repair

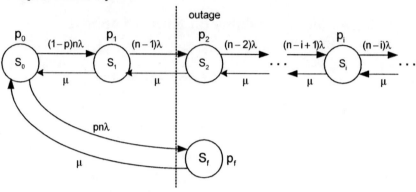

Description:

In State 0, n operational subsystems fail at a rate of λ each. (1-p) of the time, a subsystem failure causes an entry into State 1. From State 1, the failed subsystem may be repaired at a rate of μ; and State 0 is reentered. Alternatively, another operational subsystem may fail. Since there an n-1 operational subsystems in State 1, the subsystem failure rate is $(n-1)\lambda$. When a subsystem fails in State 1, State 2 is entered. In this state, there are two failed subsystems being repaired one at a time at a rate of μ. Thus, repairs occur at a rate of μ in State 2. When a subsystem is repaired, State 1 is reentered. This sequence continues for States 2 through n.

Alternatively, a subsystem failure in State 0 will experience a failover fault with a probability of p. In this case, the system enters State f, where the failed subsystem is repaired at a rate of μ; and State 0 is then reentered.

State Transitions:

State	State Transition Equation
0	$\mu p_1 + \mu p_f = n\lambda p_0$
1	$(1-p)n\lambda p_0 + \mu p_2 = [(n-1)\lambda + \mu]p_1$
2	$(n-1)\lambda p_1 + \mu p_3 = [(n-2)\lambda + \mu]p_2$
3 - n	$(n-i+1)\lambda p_{i-1} + \mu p_{i+1} = [(n-i)\lambda + \mu]p_i$

f $\quad pn\lambda p_0 = \mu p_f$

State Probabilities (in terms p_0):

State	State Probability
0	p_0
1	$(1-p)n\alpha^{-1}p_0$
2	$(1-p)n(n-1)\alpha^{-2}p_0$
3 - n	$(1-p)\dfrac{n!}{(n-i)!}\alpha^{-i}p_0$
f	$pn\alpha^{-1}p_0$
where	$\alpha = \dfrac{\mu}{\lambda}$

Failure State Probability (in terms of failure/repair rates):

State	State Probability
0	$\dfrac{\alpha^n}{D}$
1	$(1-p)n\dfrac{\alpha^{n-1}}{D}$
2	$(1-p)n(n-1)\dfrac{\alpha^{n-2}}{D}$
3 - n	$(1-p)\dfrac{n!}{(n-i)!}\dfrac{\alpha^{n-i}}{D}$
f	$p_f = pn\dfrac{\alpha^{n-1}}{D}$
where	$D = pn\alpha^{n-1} + p\alpha^n + (1-p)\sum_{0}^{n}\dfrac{n!}{(n-i)!}\alpha^{n-i}$

Dr. Bill Highleyman, Paul J. Holenstein, and Dr. Bruce Holenstein

Failure State Probability (in terms of subsystem availability a):

State	State Probability
0	$\dfrac{a^n}{D'}$
1	$(1-p)\dfrac{na^{n-1}(1-a)}{D'}$
2	$(1-p)\dfrac{n(n-1)a^{n-2}(1-a)^2}{D'}$
3 - n	$(1-p)\dfrac{\dfrac{n!}{(n-i)!}a^{n-i}(1-a)^i}{D'}$
f	$p\dfrac{na^{n-1}(1-a)}{D'}$

where $D' = (1-a)^n D$

$$= pna^{n-1}(1-a) + pa^n + (1-p)\sum_{i=0}^{n}\frac{n!}{(n-i)!}a^{n-i}(1-a)^i$$

$$= 1 - p(1-a^n) + pna^{n-1}(1-a) + (1-p)\sum_{2}^{n}(i!-1)\binom{n}{i}a^{n-i}(1-a)^i$$

$$\approx 1$$

using relation (A3-3), (1-a) << 1, and p << 1

System Failure Probability:

$$F \approx p_2 + p_f = \frac{(1-p)n(n-1)a^{n-2}(1-a)^2 + pna^{n-1}(1-a)}{D'} \approx n(n-1)(1-a)^2 + pn(1-a)$$

Simultaneous Repair

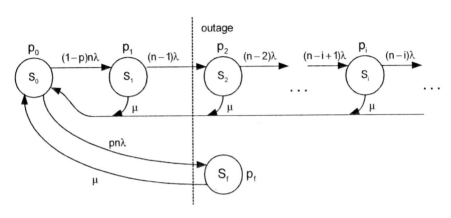

Description:

In State 0, n operational subsystems fail at a rate of λ each. (1-p) of the time, a subsystem failure causes an entry into State 1. From State 1, the failed subsystem may be repaired at a rate of μ; and State 0 is reentered. Alternatively, another operational subsystem may fail. Since there an n-1 operational subsystems in State 1, the subsystem failure rate is (n-1)λ. When a subsystem fails in State 1, State 2 is entered. In this state, there are two failed subsystems, all being repaired at the same time at a rate of μ. Thus, repairs occur at a rate of μ in State 2. When the subsystems are repaired, State 0 is reentered since all subsystems are now operational. This sequence continues for States 2 through n.

Alternatively, a subsystem failure in State 0 will experience a failover fault with a probability of p. In this case, the system enters State f, where the failed subsystem is repaired at a rate of μ; and State 0 is then reentered.

State Transitions:

State	State Transition Equation
0	$\mu\sum_{i=1}^{n} p_i + \mu p_f = n\lambda p_0$
1	$(1-p)n\lambda p_0 = [(n-1)\lambda + \mu]p_1$
2	$(n-1)\lambda p_1 = [(n-2)\lambda + \mu]p_2$
3 - n	$(n-i+1)p_{i-1} = [(n-i)\lambda + \mu]p_i$
f	$pn\lambda p_0 = \mu p_f$

State Probabilities (in terms p_0):

State	State Probability
0	p_0
1	$\dfrac{(1-p)n}{[(n-1)+\alpha]} p_0$
2	$\dfrac{(1-p)n(n-1)}{[(n-1)+\alpha][(n-2)+\alpha]} p_0$
3 - n	$\dfrac{(1-p)\dfrac{n!}{(n-i)!}}{\prod_{j=1}^{i}[(n-j)+\alpha]} p_0$
f	$pn\alpha^{-1} p_0$
where	$\alpha = \dfrac{\mu}{\lambda}$

Failure State Probability (in terms of failure/repair rates):

State	State Probability
0	α^n / D
1	$(1-p)\dfrac{n\alpha^n}{[(n-1)+\alpha]} / D$
2	$(1-p)\dfrac{n(n-1)\alpha^n}{[(n-1)+\alpha][(n-2)+\alpha]} / D$

Breaking the Availability Barrier

State	State Probability
3 - n	$(1-p)\dfrac{\dfrac{n!}{(n-i)!}\alpha^n}{\prod_{j=1}^{i}[(n-j)+\alpha]}/D$
f	$pn\alpha^{n-1}/D$

where

$$D = pn\alpha^{n-1} + \alpha^n + (1-p)\sum_{i=1}^{n}\dfrac{n!}{(n-i)!}\dfrac{\alpha^n}{\prod_{j=1}^{i}[(n-j)+\alpha]} \approx pn\alpha^{n-1} + \alpha^n$$

for $\alpha \gg 1$

Failure State Probability (in terms of subsystem availability a):

State	State Probability
0	$\dfrac{a^n}{D'}$
1	$(1-p)\dfrac{na^{n-1}(1-a)}{D'}$
2	$(1-p)\dfrac{n(n-1)a^{n-2}(1-a)^2}{D'}$
3 - n	$(1-p)\dfrac{\dfrac{n!}{(n-i)!}a^{n-i}(1-a)^i}{D'}$
f	$\dfrac{pna^{n-1}(1-a)}{D'}$

where

$$D' = (1-a)^n D = pna^{n-1}(1-a) + pa^n + (1-p)\sum_{i=0}^{n}\dfrac{n!}{(n-i)!}a^{n-i}(1-a)^i$$

$$= 1 - p(1-a^n) + pna^{n-1}(1-a) + (1-p)\sum_{2}^{n}(i!-1)\binom{n}{i}a^{n-i}(1-a)^i \approx 1$$

using $\alpha = a/(1-a)$, relation (A3-3), $(1-a) \ll 1$, and $p \ll 1$

System Failure Probability:

$$F \approx p_2 + p_f \approx \dfrac{(1-p)n(n-1)a^{n-2}(1-a)^2 + pna^{n-1}(1-a)}{D'} \approx n(n-1)(1-a)^2 + pn(1-a)$$

Dr. Bill Highleyman, Paul J. Holenstein, and Dr. Bruce Holenstein

Failover Fault Requiring System Recovery

Parallel Repair

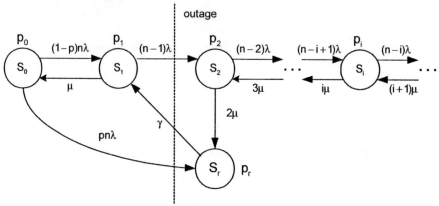

Description:

In State 0, n operational subsystems fail at a rate of λ each. (1-p) of the time, a subsystem failure causes an entry into State 1. The system continues to be operational in State 1 since only one subsystem has failed. From State 1, the failed subsystem may be repaired at a rate of μ; and State 0 is reentered. Alternatively, another operational subsystem may fail. Since there an n-1 operational subsystems in State 1, the subsystem failure rate is $(n-1)\lambda$. When a subsystem fails in State 1, State 2 is entered. At this point, the system is down since there are two failed subsystems. Each is being repaired at a rate of μ. Thus, repairs occur at a rate of 2μ in State 2. When a subsystem is repaired, the system must be recovered. Therefore, State r is entered for recovery. Recovery occurs at a rate of γ. Once recovery has been completed, State 1 is entered since there is still one subsystem down.

While in State 2, another failure can occur with a rate of $(n-1)\lambda$, in which case State 3 is entered. Repairs occur in State 3 at a rate of 3μ. Should a repair be made, State 2 is reentered. This sequence continues for States 3 through n, in which repairs are made until n-1 subsystems are operational, thus allowing system recovery to proceed.

Breaking the Availability Barrier

Alternatively, a subsystem failure in State 0 will experience a failover fault with a probability of p. In this case, the system enters State r, where the system is recovered at a rate of γ; and State 1 is then entered to await the repair of the one failed subsystem.

State Transitions:

State	State Transition Equation
0	$\mu p_1 = n\lambda p_0$
1	$(1-p)n\lambda p_0 + \gamma p_r = [(n-1)\lambda + \mu]p_1$
2	$(n-1)\lambda p_1 + 3\mu p_3 = [(n-2)\lambda + 2\mu]p_2$
3 - n	$(n-i+1)\lambda p_{i-1} + (i+1)\mu p_{i+1} = [(n-i)\lambda + i\mu]p_i$
r	$pn\lambda p_0 + 2\mu p_2 = \gamma p_r$

State Probabilities (in terms p_0):

State	State Probability
0	p_0
1	$n\alpha^{-1}p_0$
2	$\dfrac{n(n-1)}{2}\alpha^{-2}p_0$
3 - n	$\dfrac{n!}{i!(n-i)!}\alpha^{-i}p_0 = \binom{n}{i}\alpha^{-i}p_0$
r	$\dfrac{\mu}{\gamma}[n(n-1)\alpha^{-2} + pn\alpha^{-1}]p_0$
where	$\alpha = \dfrac{\mu}{\lambda}$

Failure State Probability (in terms of failure/repair rates):

State	State Probability
0	α^n / D
1	$n\alpha^{n-1} / D$
2	$\dfrac{n(n-1)}{2}\alpha^{n-2} / D$

State	State Probability
3 - n	$\binom{n}{i} \alpha^{n-i} / D$
r	$\dfrac{\mu}{\gamma}[n(n-1)\alpha^{n-2} + pn\alpha^{n-1}]/D$

where $D = \dfrac{\mu}{\gamma}[n(n-1)\alpha^{n-2} + pn\alpha^{n-1}] + \sum_{0}^{n}\binom{n}{i}\alpha^{n-i}$

Failure State Probability (in terms of subsystem availability a):

State	State Probability
0	a^n / D'
1	$na^{n-1}(1-a)/D'$
2	$\dfrac{n(n-1)}{2}a^{n-2}(1-a)^2 / D'$
3 - n	$\binom{n}{i}a^{n-i}(1-a)^i / D'$
r	$\dfrac{\mu}{\gamma}[n(n-1)a^{n-2}(1-a)^2 + pna^{n-1}(1-a)]/D'$

$D' = (1-a)^n D$

where
$$= \sum_{i=0}^{n}\binom{n}{i}a^{n-i}(1-a)^i + \dfrac{\mu}{\gamma}[pna^{n-1}(1-a) + n(n-1)a^{n-2}(1-a)^2]$$
$$= 1 + \dfrac{\mu}{\gamma}[pna^{n-1}(1-a) + n(n-1)a^{n-2}(1-a)^2] \approx 1$$

using $\alpha = a/(1-a)$, relation (A3-2), and $(1-a) \ll 1$

System Failure Probability:

$$F \approx p_2 + p_r = \frac{pn\frac{R}{r}a^{n-1}(1-a) + \frac{n(n-1)}{2}\left(\frac{r/2+R}{r/2}\right)a^{n-2}(1-a)^2]}{D'}$$

$$F \approx pn\frac{R}{r}(1-a) + \frac{n(n-1)}{2}\left(\frac{r/2+R}{r/2}\right)(1-a)^2$$

where $r = 1/\mu$ and $R = 1/\gamma$.

Dr. Bill Highleyman, Paul J. Holenstein, and Dr. Bruce Holenstein

Sequential Repair

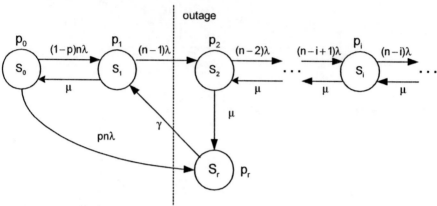

Description:

In State 0, n operational subsystems fail at a rate of λ each. (1-p) of the time, a subsystem failure causes an entry into State 1. The system continues to be operational in State 1 since only one subsystem has failed. From State 1, the failed subsystem may be repaired at a rate of μ; and State 0 is reentered. Alternatively, another operational subsystem may fail. Since there an n-1 operational subsystems in State 1, the subsystem failure rate is (n-1)λ. When a subsystem fails in State 1, State 2 is entered. At this point, the system is down since there are two failed subsystems. Subsystems are being repaired at a rate of μ. Thus, repairs occur at a rate of μ in State 2. When a subsystem is repaired, the system must be recovered. Therefore, State r is entered for recovery. Recovery occurs at a rate of γ. Once recovery has been completed, State 1 is entered since there is still one subsystem down.

While in State 2, another failure can occur with a rate of (n-1)λ, in which case State 3 is entered. Repairs occur in State 3 at a rate of μ, and should a repair be made, State 2 is reentered. This sequence continues for States 3 through n, in which repairs are made until n-1 subsystems are operational, thus allowing system recovery to proceed.

Alternatively, a subsystem failure in State 0 will experience a failover fault with a probability of p. In this case, the system enters

Breaking the Availability Barrier

State r, where the system is recovered at a rate of γ. State 1 is then entered to await the repair of the one failed subsystem.

State Transitions:

State	State Transition Equation
0	$\mu p_1 = n\lambda p_0$
1	$(1-p)n\lambda p_0 + \gamma p_r = [(n-1)\lambda + \mu]p_1$
2	$(n-1)\lambda p_1 + \mu p_3 = [(n-2)\lambda + \mu]p_2$
3 - n	$(n-i+1)\lambda p_{i-1} + \mu p_{i+1} = [(n-i)\lambda + \mu]p_i$
r	$pn\lambda p_0 + \mu p_2 = \gamma p_r$

State Probabilities (in terms p_0):

State	State Probability
0	p_0
1	$n\alpha^{-1} p_0$
2	$n(n-1)\alpha^{-2} p_0$
3 - n	$\dfrac{n!}{(n-i)!} \alpha^{-i} p_0$
r	$\dfrac{\mu}{\gamma}[n(n-1)\alpha^{-2} + pn\alpha^{-1}]p_0$

where $\alpha = \dfrac{\mu}{\lambda}$

Failure State Probability (in terms of failure/repair rates):

State	State Probability
0	α^n / D
1	$n\alpha^{n-1} / D$
2	$n(n-1)\alpha^{n-2} / D$
3 - n b	$\dfrac{n!}{(n-i)!} \alpha^{n-1} / D$
r	$\dfrac{\mu}{\gamma}[n(n-1)\alpha^{n-2} + pn\alpha^{n-1}]/D$

331

where
$$D = \frac{\mu}{\gamma}[n(n-1)\alpha^{n-2} + pn\alpha^{n-1}] + \sum_{0}^{n} \frac{n!}{(n-i)!}\alpha^{n-i}$$

Failure State Probability (in terms of subsystem availability a):

State	State Probability
0	a^n/D'
1	$na^{n-1}(1-a)/D'$
2	$n(n-1)a^{n-2}(1-a)^2/D'$
3 - n	$\frac{n!}{(n-i)!}a^{n-i}(1-a)^i/D'$
r	$\frac{\mu}{\gamma}[n(n-1)a^{n-2}(1-a)^2 + pna^{n-1}(1-a)]/D'$

$D' = (1-a)^n D$

where
$$= 1 + \frac{\mu}{\gamma}[pna^{n-1}(1-a) + n(n-1)a^{n-2}(1-a)^2] + \sum_{i=2}^{n}(i!-1)\binom{n}{i}a^{n-i}(1-a)^i$$
$$\approx 1$$

using $\alpha = a/(1-a)$, relation (A3-3), and $(1-a) \ll 1$

System Failure Probability:

$$F \approx p_2 + p_r = \frac{pn\frac{R}{r}a^{n-1}(1-a) + n(n-1)\left(\frac{r+R}{r}\right)a^{n-2}(1-a)^2]}{D'}$$

$$F \approx pn\frac{R}{r}(1-a) + n(n-1)\left(\frac{r+R}{r}\right)(1-a)^2$$

where $r = 1/\mu$ and $R = 1/\gamma$.

Simultaneous Repair

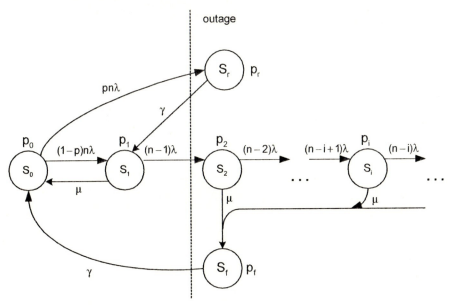

Description:

In State 0, n operational subsystems fail at a rate of λ each. (1-p) of the time, a subsystem failure causes an entry into State 1. From State 1, the failed subsystem may be repaired at a rate of μ; and State 0 is reentered. Alternatively, another operational subsystem may fail. Since there an n-1 operational subsystems in State 1, the subsystem failure rate is $(n-1)\lambda$. When a subsystem fails in State 1, State 2 is entered. This state represents a system outage. In State 2, there are two failed subsystems, all being repaired at the same time at a rate of μ. Thus, repairs occur at a rate of μ in State 2. When the subsystems are repaired, State f is entered to recover the system. The system is recovered at a rate of γ, and the system then reenters State 0 since all subsystems are now operational.

While in State 2, another failure can occur with a rate of $(n-1)\lambda$, in which case State 3 is entered. Simultaneous repairs occur in State 3 at a rate of μ; and, should a repair be made, State f is entered for system recovery. This sequence continues for States 3 through n.

Alternatively, a subsystem failure in State 0 will experience a failover fault with a probability of p. In this case, the system enters State r, where the system is recovered at a rate of γ, and State 1 is entered to await the repair of the one failed subsystem.

State Transitions:

State	State Transition Equation
0	$\mu p_1 + \gamma p_f = n\lambda p_0$
1	$(1-p)n\lambda p_0 + \gamma p_r = [(n-1)\lambda + \mu]p_1$
2	$(n-1)\lambda p_1 = [(n-2)\lambda + \mu]p_2$
3 - n	$(n-i+1)p_{i-1} = [(n-i)\lambda + \mu]p_i$
r	$pn\lambda p_0 = \gamma p_r$
f	$\mu \sum_{i=2}^{n} p_i = \gamma p_f$

State Probabilities (in terms p_0):

State	State Probability
0	p_0
1	$\dfrac{n}{[(n-1)+\alpha]} p_0$
2	$\dfrac{n(n-1)}{[(n-1)+\alpha][(n-2)+\alpha]} p_0$
3 - n	$\dfrac{\dfrac{n!}{(n-i)!}}{\prod_{j=1}^{i}[(n-j)+\alpha]} p_0$
f	$\dfrac{\mu}{\gamma} p n \alpha^{-1} p_0$
r	$\dfrac{\mu}{\gamma} \dfrac{n(n-1)\alpha^{-1}}{[(n-1)+\alpha]} p_0$
where	$\alpha = \dfrac{\mu}{\lambda}$

Failure State Probability (in terms of failure/repair rates):

State	State Probability
0	α^n / D
1	$\dfrac{n\alpha^n}{[(n-1)+\alpha]} / D$
2	$\dfrac{n(n-1)\alpha^n}{[(n-1)+\alpha][(n-2)+\alpha]} / D$
3 - n	$\dfrac{\dfrac{n!}{(n-i)!}\alpha^n}{\prod_{j=1}^{i}[(n-j)+\alpha]} / D$
f	$\dfrac{\mu}{\gamma} p n \alpha^{n-1} / D$
r	$\dfrac{\mu}{\gamma} \dfrac{n(n-1)\alpha^{n-1}}{[(n-1)+\alpha]}$

where
$$D = \frac{\mu}{\gamma} p n \alpha^{n-1} + \frac{\mu}{\gamma} \frac{n(n-1)\alpha^{n-1}}{[(n-1)+\alpha]} + \alpha^n + \sum_{i=1}^{n} \frac{n!}{(n-i)!} \frac{\alpha^n}{\prod_{j=1}^{i}[(n-j)+\alpha]}$$

$$\approx \frac{\mu}{\gamma} p n \alpha^{n-1} + \frac{\mu}{\gamma} n(n-1)\alpha^{n-2} + \sum_{0}^{n} \frac{n!}{(n-1)!} \alpha^{n-i}$$

for $\alpha \gg 1$

Failure State Probability (in terms of subsystem availability a):

State	State Probability
0	a^n / D'
1	$n a^{n-1}(1-a) / D$
2	$n(n-1) a^{n-2}(1-a)^2 / D'$
3 - n	$\dfrac{n!}{(n-1)!} a^{n-i}(1-a)^i / D'$
f	$\dfrac{\mu}{\gamma} p n a^{n-1}(1-a) / D'$

$$r \quad \frac{\mu}{\gamma}n(n-1)a^{n-2}(1-a)^2/D'$$

where $D' = (1-a)^n D$

$$\approx \frac{\mu}{\gamma}pna^{n-1}(1-a) + \frac{\mu}{\gamma}n(n-1)a^{n-2}(1-a)^2 + \sum_{i=0}^{n}\frac{n!}{(n-i)!}a^{n-i}(1-a)^i$$

$$= 1 + \frac{\mu}{\gamma}[pna^{n-1}(1-a) + n(n-1)a^{n-2}(1-a)^2] + \sum_{2}^{n}(i!-1)\binom{n}{i}a^{n-i}(1-a)^i$$

$$\approx 1$$

using $\alpha = a/(1-a)$, relation (A3-3), and $(1-a) \ll 1$

System Failure Probability:

$$F \approx p_f + p_r + p_2 \approx \frac{R}{r}pna^{n-1}(1-a)/D' + n(n-1)\left(\frac{r+R}{r}\right)a^{n-2}(1-a)^2/D'$$

$$F \approx \frac{R}{r}pn(1-a) + n(n-1)\left(\frac{r+R}{r}\right)(1-a)^2$$

where $r = 1/\mu$ and $R = 1/\gamma$.

Appendix 4 - Implementing a Data Replication Project

Overview

Implementing a data replication project to improve system availability can be easy or hard depending on the requirements and complexity of the project. But all projects, no matter how big or small, go through the same stages:

1. *Analysis of Needs.* Here you figure out your particular needs for the project.

2. *Evaluation of Options.* The good news is you usually have many options for satisfying your needs. The bad news is the same as the good news – you have many options, sometimes too many - and this can lead to general confusion. Thus, you must find, understand, and rank the various options.

3. *Resolution of Concerns.* Each implementation option usually has concerns. Can these concerns be mitigated?

4. *Implementation.* What type of implementation will be done? Is it an in-house solution, a vendor solution or some form of hybrid? Who will manage the project?

5. *Testing/Quality Assurance.* The implementation requires verification and should be put through a battery of test cases. This ensures that the implementation attains the business goals identified in the beginning and that the solution works as it was designed. The tests should range from the straightforward "Does the target data match the source, subject to any transformations?" to the strategic "Does the data available at

the target satisfy my original business goals of putting it there in the first place?"

6. *Deployment into Production.* Depending on the nature of the project, deploying data replication into production can be as easy as taking the system down, or the deployment can use "zero down time" implementation principles for performing on-line deployments.

7. *Documentation Completion.* Often overlooked, an appropriate form and amount of documentation needs to be prepared in order to describe to others how the implementation works, how to maintain it, and how to modify it to meet future business needs.

We will take a closer look at each of these stages in the rest of this appendix.

Analysis of Needs

The first step in implementing a data replication project is to assess the needs of the project. Here are some considerations that go into the needs analysis:

1. *Nature of the replication need.* Is this a mission-critical replication project (e.g., a 911 system), or is this for a low priority project (e.g., an off-line data warehouse)? How many 9s do you need (see Chapter 1)? What are the data volumes expected? Can you afford down time? What amount of data latency in the target can you afford (Chapters 3 and 4)? Do you need guaranteed replication (see Chapters 3, 4, and 5)? What are your business's particular RPO or RTO needs (see Chapter 6)?

2. *Primary Usage.* Is this a disaster recovery or business continuity solution, a heterogeneous replication feed (perhaps to feed a warehouse), or a hybrid solution? Will your applications be active on all nodes (Chapters 2, 3, 4)? Will this

Breaking the Availability Barrier

be a unidirectional feed or a bi-directional feed (Chapters 3, 4)?

3. *Hardware and operating systems.* Which of your types of hardware, LAN, and operating system(s) must be supported? Can a split system, possibly providing disaster tolerance, handle your needs (see Chapters 2 and 7)?

4. *Security.* Should the data content be protected from prying eyes, especially if it goes out over the Internet? Should the data be encrypted, and if so, how?

5. *Ease of Use.* Will you need to allocate staff to modify and monitor the complete solution? In how many sites will you need this staff? How long will it take new staff to learn the solution (See Chapter 8)?

6. *Data types.* Which of your data types must be handled (e.g. INT, CHAR, VARCHAR, NCHAR, BLOBS, CLOBS, VARARRAYS)?

7. *Data capture and apply.* Should the solution use database triggers, or should it intercept disk I/Os, perform audit trail reading, or use some other scheme to capture the data to be replicated? Will ODBC or native database calls be used to apply the data? Should the apply algorithms use pre-compiled SQL?

8. *Data cleansing, transformation, and filtering.* Is your source data dirty and in need of cleansing or transformation before loading or replication to the target? Will you be normalizing, de-normalizing, aggregating, or disaggregating the data? Will you need to merge data from a variety of sources? Will you be sending all of the source data, or perhaps just selected subsets, to the target?

9. *Target table creation and Meta Data transfer.* Do you have to maintain the target tables, or should the solution automatically create the target tables/database? As table changes are made to

the source, should these be automatically applied to the target? Should you be able to add tables or columns on-line?

10. *Initial Loading.* How should the solution load the target tables to initially synchronize the target to the source? Should it be an on-line or an off-line/bulk loading technique or perhaps a combination of the two?

11. *Database configuration.* Do you need to have your replication stream going both ways at the same time to the same database? Do you need collision or bi-directional (ping-pong) avoidance, or do you need collision identification and resolution (see Chapters 2, 3, 9, and 10)?

12. *Connectivity.* How should your replication components interact (TCP/IP, LAN, MAN, WAN, X.25, etc.)? Do you need to use a public infrastructure link (e.g., the Internet)?

13. *Cost.* What types of solutions can your company afford? Often, this is related to the cost of down time, so you should calculate that cost prior to proceeding. Some companies have added hidden costs to down time calculations. They include lost customers, damage to reputation, fines, and even stock devaluation.

It is recommended that a data flow diagram or architecture diagram be created for the intended solution. Convene a meeting among the interested parties at your company to critique the proposed solution. Additional needs are likely to emerge. Make sure you have management approval and support for your needs, especially any business continuity needs identified.

Usually, the outcome of this stage for larger companies is a formal requirements specification. It is not unusual for larger companies to insist that a nondisclosure agreement be signed before the requirements specification is provided to a vendor.

Breaking the Availability Barrier

If you do not produce a formal requirements specification, you should at least make a complete list of your needs before going into the Evaluation of Options stage.

Evaluation of Options

Generally, there are two primary options, usually characterized as "make vs. buy":

1. *Building a custom solution.* Either in-house staff or contractors design and implement the solution.

2. *Buying a complete or semi-complete solution.* A vendor is selected, one that meets most, if not all, of the requirements of the project. In-house staff, vendor professional services, or contractors tailor the vendor's product to fit the needs.

Building a custom solution

Often, companies have a highly trained staff that is able to implement complex projects. Sometimes, companies have access to a trusted group of computer software contractors. In such cases, a custom solution implementation may be the best way to satisfy the requirements. Often, the initial high cost of a purchased solution prompts companies to build a custom solution. Wise companies look at the total cost of the solution over many years (5+) before making a decision since maintenance and enhancements costs can often dwarf the initial implementation costs.

Also, be wary of in-house staff or contractors underestimating a project. Usually, programmers and contractors are optimistic; and they want to work on something new, cool, or lucrative. Many custom projects fail because of unforeseen complications that delay the project and drive up the costs. Ask yourself – "What business am I really in?" If it is not the nuts and bolts of the data replication business, consider partnering with a qualified vendor.

Where possible, use well-experienced and certified professionals. Some relevant professional types for your project may include

Dr. Bill Highleyman, Paul J. Holenstein, and Dr. Bruce Holenstein

Microsoft Certified Software or System Engineers (www.microsoft.com/mcp), Certified Business Continuity Planners (http://www.drii.org), HP Accredited System Engineers (http://www.hp.com/certification), and HP Mission-Critical Consultants (http://h71033.www7.hp.com/object/mcprogde.html).

Buying a complete or semi-complete solution

A purchased solution can reduce the risk of the project implementation because the development problems have been worked out ahead of time and because the acquisition costs are well documented. In most cases, it is wise to avoid beta and "Release 1.0" products if possible. The reason to avoid unproven software is that vendor contracts always disclaim warranties, for example, if their software corrupts your database or causes you to lose money or time.

COTS Data Replication Products

Many vendors today manufacture *commodity* or *commercial off-the-shelf (COTS)* data replication products. Most products are geared to very specific purposes, but the maturity of the product and speed of innovation and implementation is high. Any list of data replication products is obsolete nearly as soon as it is drawn. The best way to find current information on existing products is to do a web search with search terms such as the following:

1. Data replication, synchronization, vaulting
2. Remote mirroring
3. Bi-directional, ping-pong, looping, circular replication avoidance
4. Multi-Master replication
5. Peer-to-peer replication
6. Asynchronous and synchronous replication
7. Heterogeneous and homogeneous replication
8. High-availability products
9. Business continuity products

You will get the best results if you include the particular platforms and operating systems of interest in the search. Some companies to consider are listed at the end of this appendix.

Distinguishing Criteria

After building a list of candidate companies and products, how do you distinguish between the various choices? Compare your needs with the capabilities offered by the vendors. In particular, pay attention to the following:

1. *Review web site and associated white papers.* Be particularly sensitive to the materials that are unclear or imply that a particular feature may not be available until the future. If unsure, ask. Often, "vaporware" never gets built.

2. *Conduct a phone and/or email interview.* At this point, you are attempting to make sure that the product specifications and price fit your needs. You are not yet attempting to negotiate a final price or acquisition terms.

3. *Conduct a trial of the software.* Be particularly wary if the vendor attempts to charge a fee for the trial. Don't necessarily pick the first (eventually) successful trial software – some vendors will wear down a potential customer by having time consuming trials so that there may not be time left to finish your evaluations. You should build at a minimum a best alternative to your first choice.

4. *Evaluate the pricing.* Does the vendor give you a guarantee that its solution will work? Can you get out of a long-term contract? Check future maintenance cost obligations.

5. *Training.* How will your staff come up to speed on the solution? Does the vendor offer on-site or off-site formal training?

6. *Customer Support.* What forms of support and support contracts does the vendor offer (e.g., phone, e-mail, fax, etc)?

What are their hours of operations for the time zones in which you are interested? Is 24x7 coverage available; and if so, at what cost? What service level agreement (SLA) interests you (e.g., 1 hour response, 4 hour response, etc)?

7. *Other Considerations.* How long has the company been in business? What is their willingness to work with you on your needs and to discuss your complex problems? Do they offer a method for you to "influence" future development for highly desired new features?

8. *Talk to references.* Are there other companies like your own that are successfully using the product for your intended purpose?

Resolution of Concerns

Often, the choice is not clear as to the best course of action. For instance, you wind up having three possible options, each of which has "flaws." You may like Solution A, but the company has no track record of happy customers; and the trial software crashed. Solution B may not handle a particular data type, and the company will not give a warranty. Solution C, your build-it-internally option, will be undertaken by some programmers who had significant cost and schedule overruns on their last project.

Making these complex decisions is difficult. The best strategy the authors have found is to openly communicate the concerns with the interested parties and to get their assurances (in writing if appropriate) that there will not be a problem. Document and distribute the interaction to show due diligence in case you have to demand performance at a later time or need to reconstruct responsibilities should the project fail in some horrible way.

Project Implementation

Implementing a data replication project is no different from any other complex project, and so the same project management formalisms should be used. In any case, here are some considerations for implementing a replication project:

1. *Timeframe.* Every project should have a project plan with explicit goals that outlines the steps that will be taken to implement the project. Keep in mind that programmers are usually optimistic; therefore, slack must be built into the schedule. Often, the testing process takes more time than the initial development process.

2. *Manpower.* Select experienced and certified personnel whenever possible. One man-month of an experienced developer's time is often worth many months of a new hire's time.

3. *Management.* The same thing applies to management of the project. The manager should be technical enough to understand the root causes of the schedule slips and when to resist "creeping elegance" (unnecessary new features being added).

4. *Operations.* Don't forget to talk to the computer systems and operations staff early and often, and address their concerns. They are the ones who care about disk space, CPU utilization, loading factors, configuration changes, and other systems management and operational concerns.

5. *Deployment.* Will the production deployment be a "big bang" implementation over a weekend; or is the deployment possibly transitional, one application at a time? Can source application down time be tolerated, or must this be a zero down time

deployment? In case of failure, is a fallback strategy needed, and if so what is it?

6. *Isolation and testing.* A separate development and test system is a wise investment for most companies. Usually, vendors are not able to simulate your complex environment in their product QA cycle. So you must plan to test all significant operational aspects of the vendor's release *each* time one is received. A formal Verification and Validation (V&V) process is often a good investment as it will help you verify that your original business objectives are fully met by the final solution and will provide the blue print for making sure this is so.

7. *Documentation.* Always ensure that you have full and complete documentation for the solution. This will include all that provided by your chosen vendor plus any that is internally developed. Make sure that this includes overall project architecture information (so those that come along later can understand why and how the project was undertaken), specific technical documentation on the products used or developed (and how to maintain them), and complete operational procedures for maintaining the solution in day-to-day operations. Generally, even though you may have purchased a vendor solution, you will need to invest in generating the specific procedures for your environment.

Company Listing

The following is a list of many companies making high availability data replication products as of the date of publication of this text. Start with the web site addresses below for their latest information. If a web page address is no longer valid, attempt to do a web search on the product (in italics) or company name. The bracketed information was compiled as of this book printing as the primary relevant use of the product. However, it may not be complete or accurate, so check for yourself. All companies and products mentioned are trademarks of their respective holders. When the acronym DR is used below, it means "disaster recovery." EAI means

Breaking the Availability Barrier

"enterprise application integration." ETL means "extraction, transformation, and loading."

Data Junction Corp., *Data Junction*, www.datajunction.com [DR, Data Warehousing, Extraction Transformation Loading (ETL)]

Data Mirror Corp., *IReflect*, www.datamirror.com [DR]

EMC Corporation, *Clariion, and many others*, www.emc.com [SAN, mirroring]

GoldenGate, Inc. *Global Synchronization* and *Extractor/Replicator*, www.goldengate.com [asynchronous DR replication on HP NonStop Servers (Tandem NSK), Unix, Windows and others]

Hewlett Packard Company, *RDF*, www.hp.com [asynchronous DR replication exclusively for HP NonStop Servers (Tandem NSK)]

ITI Shadowbase, a division of Gravic, Inc., *Shadowbase*, www.iticsc.com [asynchronous and synchronous DR replication on HP NonStop Servers (Tandem NSK), Unix, Windows, and others; bi-directional active/active replication]

Lakeview Technology, *OmniReplicator*, www.lakeviewtech.com [DR, bi-directional replication, Business Continuity]

Network Technologies International, Inc., *DR Net*, www.networktech.com [asynchronous DR replication, primarily for HP NonStop Servers (Tandem NSK)]

Oracle Corporation, *Data Guard, Advanced Replication, Streams, Real Application Clusters, and others*. www.oracle.com [extensive replication offerings, primarily for Oracle platforms]

Parallel DB Corp., *PdbReplicate*, www.paralleldb.com [DR]

Dr. Bill Highleyman, Paul J. Holenstein, and Dr. Bruce Holenstein

Peer Direct Corporation, *Peer Direct Enterprise*, www.peerdirect.com [DR, EAI]

Pervasive Software, *DataExchange*, www.pervasive.com [DR, data warehousing]

Quest Software, *SharePlex,* www.quest.com [replication primarily for Oracle platforms]

Sybase, Inc., *Sybase Replication Server*, www.sybase.com [replication primarily for Sybase platforms]

Veritas Software, *Storage Replicator, Volume Replicator and others*, www.veritas.com [extensive offering, primarily mirroring]

And many more ...

References and Suggested Reading

The following references include those mentioned in the various chapters in this book. They also include some additional references that are quite useful for further background material.

Advanced Computer and Network Corporation, "*RAID 6*," www.acnc.com.

Barker, R.; Massiglia, P.; Storage Area Network Essentials, Wiley Computer Publishing; 2002.

Bartlett, W.; "*Indestructible Scalable Computing*," ITUG Summit presentation; September, 2001.

Buckle, R.; Highleyman, W. H.; "The New NonStop Advanced Architecture: A Massive Jump in Processor Reliability," The Connection, Volume 24, Number 6; September/October, 2003.

Compaq Computers, "*Disaster Tolerance: The Technology of Business Continuity*," www.techguide.com.

Devraj, V. S.; "*Oracle 24x7 Tips & Techniques*," Oracle Press; 2000.

Einhorn, S. J.; "*Reliability Prediction for Repairable Redundant Systems*," Proceedings of the IEEE; February, 1963.

Faithful, M.; "*The a-BCP of DR*," The Connection, Issue 23, Number 6; November/December, 2002.

Fox, A.; Patterson, D.; "*Self-Repairing Computers*," Scientific American; June, 2003.

Dr. Bill Highleyman, Paul J. Holenstein, and Dr. Bruce Holenstein

Gray, J.; *"Why Do Computers Stop and What Can We Do About It?"* 5th Symposium on Reliability in Distributed Software and Database Systems; 1986.

Gray, J.; Reuter, A.; Transaction Processing: Concepts and Techniques, Morgan Kaufman; 1993.

Gray, J.; Helland, P.; O'Neil, P.; Shasha, D.; *"The Dangers of Replication and a Solution,"* ACM SIGMOD Record (Proceedings of the 1996 ACM SIGMOD International Conference on Management of Data), Volume 25, Issue 2; June, 1996.

Hewlett-Packard, *"HP NonStop servers ranked tops in availability by Standish, beating IBM Sysplex and RS/6000 Non-Clusters,"* NonStop Computing websites; September, 2002.

Highleyman, W.H.; *"Distributing OLTP Data Via Replication,"* The Connection, Volume 16, No. 2; April/May, 1995.

Highleyman, W. H.; *"Reliability Analysis – It's More Important Than Ever,"* The Connection, Volume 22, Number 3; May/June, 2001.

Highleyman, W. H.; *"The Impact of Mean Time to Repair on System Availability,"* ITI, Inc. White Paper, August 19, 2002.

Highleyman, W. H.; Holenstein, P. J.; Holenstein, B. D.; Six Part Availability Series, The Connection, beginning Volume 23, No. 1; September/October, 2002.

Highleyman, W. H.; Holenstein, P. J.; Holenstein, B. D.; *"Method of Increasing System Availability by Splitting a System,"* United States Patent Application 10/368,315; February 13, 2003.

Highleyman, W. H.; Holenstein, P. J.; Holenstein, B. D.; *"Method of Increasing System Availability by Assigning Process Pairs to Processor Pairs,"* United States Patent Application 10/367,675; February 13, 2003.

Holenstein, B. D.; Waterstraat, M.; "*Integrating Platforms in the E-Commerce Enterprise*," The Connection, Volume 21, Number 4; July/August, 2000.

Holenstein, P. J.; Holenstein, B. H.; "*High-Availability Web-site/NSK Cooperative Processing Using Database Replication in a ZLE Architecture*," The Connection, Volume 22, Number 5; September/October, 2001.

Holenstein, P. J.; Holenstein, B. D.; Strickler, G. E.; "*Synchronization of Plural Databases in a Database Replication System,*" United States Patent application 20030037029; August 15, 2001.

Holenstein, B. D.; Holenstein, P. J.; Strickler, G. E.; "*Collision Avoidance in Data Replication Systems*," United States Patent Application No. 20020133507; Sept. 19, 2002.

Knapp, H. W.; "*The Natural Flow of Transactions*," ITI, Inc. White Paper; 1996.

LaPedis, R.; "*Developing Contingency Plans for the Recovery of Critical Data and Applications*," Compaq White Paper; 1999.

LaPedis, R.; "*RTO and RPO Not Tightly Coupled*," Disaster Recovery Journal; Summer, 2002.

LaPedis, R.; "*Will Enterprise Storage Replace NonStop RDF?*" The Connection, Volume 23, Issue 6; November/December, 2002.

Liebowitz, B. H.; Carson, J. H.; Chapter 8 – "*Reliability Calculations*," Multiple Processing Systems for Real-Time Applications, Prentice-Hall; 1985.

Marcus, E.; Stern, H.; Blueprints for High Availability: Designing Resilient Distributed Systems, Wiley Computer Publishing; 2000.

Pfister, G. F.; In Search of Clusters: The Ongoing Battle In Lowly Parallel Computing, Prentice-Hall PTR; 1998.

Dr. Bill Highleyman, Paul J. Holenstein, and Dr. Bruce Holenstein

Shannon, T. C.; *"HP Brings Enterprise-Class Storage Capabilities to the Midrange,"* Shannon Knows HPC, Volume 10, Issue 18; April 15, 2003.

The Standish Group, *"The New High Availability – A Non Stop Continuous Processing Architecture (CPA),"* Standish Group Research Note; 2002.

The Standish Group, VirtualBEACON, Issue 244; September, 2002.

Strickler, G. W.; Knapp, H. W.; Holenstein, B. D.; Holenstein, P. J.; *"Bi-directional Database Replication Scheme for Controlling Ping-Ponging,"* United States Patent 6,122,630; Sept. 19, 2000.

Thornburgh, R. H.; Schoenborn, B. J.; Storage Area Networks, Prentice-Hall PTR; 2001.

Weygant, P. S.; Clusters for High Availability, Prentice-Hall PTR; 2001.

Wong, B. L.; Configuration and Capacity Planning for Solaris Servers, Prentice-Hall.

Wood, A.; *"Availability Modeling,"* Circuits and Devices; May, 1994.

Wood, A; *"Predicting Client/Server Availability,"* Computer, Volume 28, Number 4; April, 1995.

Index

ACID (atomic, consistent, independent, durable), 182, 228, 229, 237
Active/active, 29, 31, 38, 49, 51, 73, 75, 78, 81, 134, 136, 141, 145, 146, 147, 150, 152, 160, 162, 169, 172, 187, 218, 222, 225, 234, 242, 262, 347
Advanced Replication, Streams, 347
Amtrak, 24, 25
Asynchronous, xix, xx, xxiii, 44, 46, 55, 56, 59, 60, 62, 66, 68, 75, 76, 77, 86, 88, 103, 104, 105, 136, 141, 144, 145, 146, 147, 149, 162, 170, 176, 178, 183, 185, 186, 187, 188, 190, 200, 202, 203, 204, 205, 206, 207, 208, 212, 213, 214, 215, 222, 224, 225, 226, 229, 242, 243, 262, 265, 294, 298, 347
Availability, xxi, xxiii, xxv, 5, 6, 14, 20, 24, 27, 30, 34, 46, 49, 108, 117, 118, 156, 158, 168, 172, 175, 271, 278, 280, 281, 293, 294, 301, 350, 351, 352
approximation, xxiii, 301
backup, 10, 14, 16, 29, 37, 49, 61, 72, 73, 78, 109, 113, 114, 115, 123, 126, 128, 129, 130, 133, 134, 137, 139, 140, 142, 143, 144, 147, 150, 151, 152, 175, 217, 218, 220, 221, 222, 234, 241, 272
degradation, 11
mission critical, 24, 151, 338
nines, 7
redundancy, 8, 9, 10, 22, 349
redundant, xx, xxiv, 3, 8, 9, 10, 11, 16, 21, 24, 25, 27, 28, 45, 105, 108, 109, 156, 163, 167, 168, 171, 271, 272, 305
reliability, 3, 5, 15, 22, 34, 40, 110, 349, 350, 351
sparing, 15
Backup, 10, 14, 16, 29, 37, 49, 61, 72, 73, 78, 109, 113, 114, 115, 123, 126, 128, 129, 130, 133, 134, 137, 139, 140, 142, 143, 144, 147, 150, 151, 152, 175, 217, 218, 220, 221, 222, 234, 241, 272
magnetic tape, 129, 130, 143, 221, 222
Bartlett, W., xxv, 37, 112, 151, 177, 349
Buckle, R., xxv, 34, 349

Carson, John H., xxv, 15, 351
Change queue, 58, 59, 60, 62, 65, 74, 104, 238, 241, 246, 248, 249, 250, 251, 258, 265, 266
  audit trail, 130, 133, 139, 142, 236, 244, 251, 258, 339
  change log, 258
  Database of Change (DOC), 58, 230, 238, 240, 241, 248, 257, 258, 260, 266
Clariion, 347
Clustering
  Application Clustering Services (ACS), 171
Collisions, xix, xx, xxii, 38, 42, 51, 66, 67, 68, 69, 70, 71, 72, 74, 75, 76, 77, 79, 80, 81, 103, 104, 135, 136, 142, 145, 146, 152, 161, 162, 163, 169, 172, 176, 178, 181, 184, 187, 188, 190, 191, 197, 198, 199, 200, 202, 203, 204, 205, 206, 207, 209, 212, 213, 215, 224, 226, 229, 232, 242, 262, 294, 298, 340
  avoidance, 39, 81, 207, 208, 225, 340, 342
    partitioning, 67
    synchronous replication, 39, 56, 63, 68, 81, 104, 136, 163, 182, 183, 187, 223
  detection, 68
    versioning, 69
  hot spot, 188, 190, 191, 213
  resolution, 70
    fuzzy replication, 54, 71
    generic algorithms, 69
    manual, 72
Communication channel
  efficiency, 93
  propagation time, 84, 88, 89, 91, 96, 97, 98, 136, 177, 294
Coordinated commits, xx, 82, 84, 86, 87, 89, 90, 91, 93, 94, 95, 97, 98, 101, 102, 103, 137, 143, 152, 157, 162, 166, 170, 177, 182, 188, 191, 202, 203, 204, 205, 209, 214, 293
  ready to synchronize (RTS) token, 223
  ready-to-commit (RTC) token, 85, 88, 89, 94, 98, 104, 223
Data Guard, 347
Data integrity, 61
Data Junction Corp., 347
Data Mirror Corp, 347
Data replication, 42, 48, 230, 342
  active/active, 29, 31, 38, 49, 51, 73, 75, 78, 81, 134, 136, 141, 145, 146, 147, 150, 152, 160, 162, 169, 172, 187, 218, 222, 225, 234, 242, 262, 347
  active/passive, 29
  asynchronous, xix, xx, xxiii, 44, 46, 55, 56, 59, 60, 62, 66, 68, 75, 76, 77, 86, 88, 103, 104, 105, 136, 141, 144, 145, 146, 147, 149, 162, 170, 176, 178, 183,

185, 186, 187, 188, 190,
200, 202, 203, 204, 205,
206, 207, 208, 212, 213,
214, 215, 222, 224, 225,
226, 229, 242, 243, 262,
265, 294, 298, 347
data loss, 62, 141
database corruption, 63
functional segmentation,
   49
latency, 37, 38, 42, 53,
   56, 61, 62, 63, 66, 74,
   75, 76, 77, 78, 80, 86,
   89, 91, 94, 104, 142,
   144, 146, 147, 152,
   170, 176, 183, 186,
   187, 189, 190, 197,
   201, 202, 204, 205,
   213, 222, 225, 226,
   232, 240, 293
no performance penalty,
   60
replication latency, 37,
   38, 42, 53, 56, 61, 62,
   63, 66, 74, 75, 76, 77,
   78, 80, 86, 89, 91, 94,
   104, 142, 144, 146,
   147, 152, 170, 176,
   183, 186, 187, 189,
   190, 197, 201, 202,
   204, 205, 213, 222,
   225, 226, 232, 240, 293
bi-directional, 38, 65, 135,
   225, 342, 352
campus environments, 97,
   136, 163, 170
collisions, 66, 67, 68, 69, 75,
   80, 81, 137, 184, 186,
187, 190, 197, 203, 204,
   223, 351
conflicts, xxii, 51, 66, 80,
   181, 197, 298
data manipulation, 60
directionality, 49, 50
disaster tolerance, 46, 349
distributed application, 45
distributed system, 96, 351
engine, xxiii, xxvi, 47, 48,
   51, 52, 53, 54, 55, 56, 58,
   59, 60, 61, 62, 63, 64, 65,
   66, 69, 71, 74, 81, 86, 93,
   96, 134, 135, 143, 148,
   185, 202, 204, 206, 207,
   209, 217, 222, 229, 230,
   232, 233, 238, 239, 240,
   241, 242, 243, 248, 260,
   261, 262, 263, 265, 267,
   268
applier, 57, 60, 65, 69, 74,
   75, 229, 232, 237, 245,
   248, 249, 250, 252,
   254, 256, 257, 258,
   259, 260, 261, 262,
   263, 265, 266, 267, 268
audit trail, 130, 133, 139,
   142, 236, 244, 251,
   258, 339
change queue, 58, 59, 60,
   62, 65, 74, 104, 238,
   241, 246, 248, 249,
   250, 251, 258, 265, 266
communication queue, 58
DOC, 58, 230, 240
end-to-end threads, 246
expected ends, 246
extractor, 57, 60, 65, 74,
   229, 230, 232, 237,

245, 246, 248, 249, 250, 251, 252, 254, 261, 265, 267, 268, 347
multi-threaded, 51, 58, 243, 244, 252, 254, 257, 258, 259, 260, 265, 266, 268
physical replication, 224, 225
queuing, 52, 53
replay queue, 59, 63
serializer, 246, 248, 249
single-threaded, 238, 240
target updating, 53
Enterprise Application Integration (EAI), 47
geographically distributed, 96
heterogeneous, 47, 60, 342
highly available, 46, 61
highly scalable, xxiv, 10, 93, 162
highly secure, 62
latency
  collisions, xix, xx, xxii, 38, 42, 51, 66, 67, 68, 69, 70, 71, 72, 74, 75, 76, 77, 79, 80, 81, 103, 104, 135, 136, 142, 145, 146, 152, 161, 162, 163, 169, 172, 176, 178, 181, 184, 187, 188, 190, 191, 197, 198, 199, 200, 202, 203, 204, 205, 206, 207, 209, 212, 213, 215, 224, 226, 229, 232, 242, 262, 294, 298, 340

localization, 46
multi-node, 150, 185, 201, 230
network storage, 165, 166, 171
non-invasive, 60
non-partitioned, 49, 51
partitioned, 39, 49, 51, 65, 67, 135, 140, 208
ping-ponging, 38, 64, 65, 135, 225, 352
products, 342
redundancy, 8, 9, 10, 22, 349
synchronous, xx, 39, 51, 54, 55, 56, 63, 68, 76, 77, 81, 89, 92, 104, 136, 141, 142, 144, 163, 169, 181, 182, 183, 187, 189, 191, 201, 205, 209, 211, 214, 223, 295, 347, 351
  application latency, 56, 76, 86, 87, 89, 91, 93, 95, 97, 98, 99, 101, 137, 144, 183, 224, 293
  coordinated commits, xx, 82, 84, 86, 87, 89, 90, 91, 93, 94, 95, 97, 98, 101, 102, 103, 137, 143, 152, 157, 162, 166, 170, 177, 182, 188, 191, 202, 203, 204, 205, 209, 214, 293
  deadlocks, xxii, 51, 102, 178, 181, 184, 185, 186, 187, 189, 191, 193, 203, 205, 206, 207, 208, 213, 214, 215, 256, 258, 262, 268

dual writes, 82, 84, 85,
  86, 87, 90, 91, 93, 94,
  95, 97, 98, 99, 101,
  102, 157, 162, 163,
  170, 177, 182, 188,
  191, 202, 203, 204,
  205, 214, 293
 efficiency, 47, 53, 59, 86,
  89, 90, 91, 93, 95, 96,
  97, 98, 100, 101, 147,
  209, 232, 293
 system maintenance, 47
 system migration, 48
 unidirectional, 49, 72, 73,
  133, 139, 140, 144
Data warehousing, 47, 347
Database
 non-partitioned, 49, 51
 RAID, 168, 169, 171, 228,
  349
 striping, 169
Database of Change (DOC),
  58, 230, 238, 240, 241, 248,
  257, 258, 260, 266
DataExchange, 348
Deadlocks, xxii, 51, 102, 178,
  181, 184, 185, 186, 187,
  189, 191, 193, 203, 205,
  206, 207, 208, 213, 214,
  215, 256, 258, 262, 268
 hung transaction, 143
 intelligent locking protocol
  (ILP), 95, 102, 103, 186,
  193, 196, 201, 206, 207,
  208
 lock latency, 178, 186, 189,
  193, 194, 201, 202, 203,
  204, 205, 206, 207, 208,
  214, 298

mutex, 102, 103, 207
mutual waits, 185, 189, 193,
  197, 201, 206, 298
resolution
 asynchronous database
  access, 265
 base level audit, 263
 lock table, 209
 master node, 69, 103,
  178, 208
 timeouts, 186
Directionality, 49, 50
 unidirectional replication,
  49, 72, 73, 133, 139, 140,
  144
Disaster tolerance, 46, 127,
  349, 351
Disk farms, 167
Distributed application, 45
Distributed system, 96, 351
Down time, 4, 123
 restore time, 122, 123
DR Net, 347
Dual writes, 82, 84, 85, 86, 87,
  90, 91, 93, 94, 95, 97, 98,
  99, 101, 102, 157, 162, 163,
  170, 177, 182, 188, 191,
  202, 203, 204, 205, 214,
  293, 296
 round trip operations, 90,
  91, 97, 99, 136, 293, 294,
  296
 serial, 101, 296
Duplicate transactions, 75
Einhorn, S. J., 22, 349
EMC Corporation, 347
Enterprise Application
  Integration (EAI), 47

Operational Data Store (ODS), 47
Enterprise storage, 126, 351
Extractor/Replicator, 347
Failover, xxiii, xxiv, 43, 72, 73, 105, 113, 115, 116, 117, 118, 119, 120, 121, 152, 158, 175, 177, 234, 271, 272, 275, 276, 277, 279, 280, 281, 282, 283, 284, 297, 305, 306, 317, 320, 323, 327, 330, 334
   failover fault, xxiii, xxiv, 43, 113, 115, 117, 118, 119, 120, 121, 152, 158, 175, 177, 271, 272, 275, 276, 277, 279, 280, 281, 282, 283, 284, 297, 305, 306, 317, 320, 323, 327, 330, 334
   network failure, 73, 145
   recovery, 74, 104
   source node failure, 72
   target node failure, 73
Failover fault, xxiii, 43, 117, 118, 119, 120, 121, 152, 158, 175, 177, 271, 272, 279, 281, 283, 284, 305
   scenarios
      no failover faults, 283
Failure
   fault, xxiii, 305
   modes, 16, 17, 20
   network failure, 73, 74, 75, 81, 103, 104, 111, 125, 225, 230, 249
   recovery, 73, 74, 75, 103, 105, 116, 125
   source node failure, 72

system failure, 11, 13, 18, 29, 31, 36, 77, 78, 107, 111, 117, 121, 139, 140, 150, 166, 218, 222, 229, 275, 277, 289, 293, 302, 303
   target node failure, 73
Failure modes, 16, 17, 20
Fault tolerance, 110, 113, 123
   redundant, xx, xxiv, 3, 8, 9, 10, 11, 16, 21, 24, 25, 27, 28, 45, 105, 108, 109, 156, 163, 167, 168, 171, 271, 272, 305
Fox, A., 121, 349
Fuzzy replication, 54, 71
Gartner Group, 8
Global Synchronization, 347
Golden rule, 121
GoldenGate, 347
Gravic, xxv, xxvi, 347
Gray, Jim, xxv, 5, 40, 55, 56, 68, 80, 109, 110, 111, 162, 176, 182, 187, 188, 193, 197, 199, 205, 228, 350
Hewlett Packard Company, 347
Highleyman, W. H., xix, xxv, xxvii, 34, 349, 350
Holenstein, Bruce D., 81, 137, 223, 350, 351, 352
Holenstein, Paul J., 350, 351, 352
HP, xvii, xxiv, xxv, xxvi, 4, 8, 11, 34, 37, 40, 111, 112, 114, 123, 170, 171, 342, 347, 350, 352

Intelligent locking protocol (ILP), 95, 102, 103, 186, 193, 196, 201, 206, 207, 208
IReflect, 347
ITI Shadowbase, xxvi, 347
Knapp, H. W., xxv, 232, 351, 352
Lakeview Technology, 347
Lapedis, R., xxv, 126, 127, 351
Latency, 86, 186, 191, 202, 211
  application, 56, 76, 86, 87, 89, 91, 93, 95, 97, 98, 99, 101, 137, 144, 183, 224, 293
Liebowitz, Burt H., xxv, 15, 351
Mainframes, 8
Master node, 69, 103, 178, 208
Mean Time Before Failure (MTBF), xxiii, 4, 5, 7, 18, 19, 23, 36, 40, 41, 78, 108, 124, 156, 159, 161, 164, 167, 168, 176, 273, 293, 306
Mean Time To Repair (MTR), xxiii, 4, 5, 7, 17, 18, 19, 23, 36, 40, 108, 110, 116, 122, 124, 156, 161, 176, 273, 274, 276, 293, 294, 306
Mirrored disks, 40, 41, 163, 165, 166, 167, 171, 342, 347, 348
  split, 165
Mirroring
  disaster recovery, 127, 351
  enterprise storage, 126, 351
Multi-threading, 51, 243
Natural flow, 232, 351
  restart scenarios, 245

Network storage, 165, 166, 171
Network Technologies International, Inc., 347
Niehaus, Carl, xxv, 114, 177
Nines (9s), xvii, xviii, xxiii, 1, 3, 6, 7, 10, 15, 16, 19, 21, 22, 23, 24, 27, 33, 40, 42, 78, 108, 109, 110, 117, 123, 155, 161, 164, 165, 168, 169, 175, 176, 220, 338
Node, 39, 72, 73, 79, 150, 210, 218
  database, 165, 182, 183, 187, 188, 190, 193, 197, 210, 211, 213
  processing nodes, 45, 46, 48, 166, 167, 188, 210, 211, 218
NonStop, xvii, xx, xxiv, xxvi, 8, 11, 24, 34, 37, 40, 49, 87, 111, 123, 126, 170, 347, 349, 350, 351
  HP, xvii, xxiv, xxv, xxvi, 4, 8, 11, 34, 37, 40, 111, 112, 114, 123, 170, 171, 342, 347, 350, 352
  Indestructible Scalable Computing, 37, 349
  ServerNet, 170, 171
OmniReplicator, 347
Operational Data Store (ODS), 47
Oracle Corporation, 347
Outage, xvii, xviii, xx, 4, 12, 28, 35, 36, 42, 103, 104, 108, 109, 110, 111, 112, 113, 114, 115, 116, 117, 118, 120, 121, 122, 123,

127, 130, 142, 145, 146, 151, 155, 156, 157, 159, 161, 167, 171, 176, 177, 181, 271, 272, 273, 274, 275, 276, 277, 279, 285, 287, 306, 333

failover fault, xxiii, xxiv, 43, 113, 115, 117, 118, 119, 120, 121, 152, 158, 175, 177, 271, 272, 275, 276, 277, 279, 280, 281, 282, 283, 284, 297, 305, 306, 317, 320, 323, 327, 330, 334

failure, xvii, xviii, xx, xxiii, 4, 5, 7, 11, 12, 13, 14, 15, 16, 17, 18, 19, 21, 22, 23, 25, 28, 29, 30, 31, 32, 33, 34, 35, 36, 37, 38, 40, 42, 43, 44, 46, 52, 53, 58, 61, 62, 72, 73, 74, 75, 76, 77, 78, 83, 85, 86, 103, 104, 105, 107, 108, 109, 112, 113, 114, 115, 116, 117, 118, 119, 120, 121, 124, 125, 126, 130, 133, 139, 140, 141, 144, 146, 147, 150, 151, 152, 155, 156, 157, 158, 160, 161, 163, 164, 165, 166, 169, 171, 175, 176, 177, 181, 217, 218, 221, 222, 225, 228, 229, 230, 249, 271, 272, 273, 274, 275, 276, 277, 279, 281, 282, 283, 284, 287, 288, 289, 293, 302, 303, 305, 306, 308, 309, 311, 312, 314, 315, 317, 318, 320, 321, 323, 324, 326, 327, 330, 331, 333, 334, 335, 346

trigger, 52, 82, 84, 113, 114, 230, 339

fault, xxiii, xxiv, xxvi, 3, 10, 11, 12, 16, 21, 23, 27, 37, 40, 61, 72, 109, 113, 114, 115, 117, 118, 119, 120, 123, 143, 158, 164, 172, 181, 224, 271, 272, 275, 276, 277, 280, 281, 282, 283, 284, 285, 297, 305, 306, 317, 320, 323, 327, 330, 334

human error, xx, 25, 43, 111, 112, 120, 155, 158, 285

independent failures, 115

infant mortality, 110

recovery, xx, xxiii, xxvi, 15, 41, 49, 60, 61, 97, 103, 105, 107, 112, 113, 116, 117, 118, 119, 120, 121, 122, 123, 124, 125, 126, 128, 129, 130, 133, 134, 137, 139, 140, 141, 143, 144, 145, 146, 149, 150, 151, 152, 155, 156, 158, 175, 177, 218, 221, 222, 249, 271, 272, 275, 276, 277, 279, 280, 282, 283, 284, 285, 305, 306, 326, 330, 333, 338, 346

repair, xxiii, 4, 17, 18, 23, 36, 40, 105, 107, 108, 112, 113, 116, 117, 118, 120, 121, 122, 124, 125, 155, 156, 157, 164, 175, 177, 271, 272, 273, 274,

275, 276, 277, 278, 280,
281, 282, 283, 284, 285,
287, 288, 293, 294, 297,
305, 306, 309, 312, 315,
318, 321, 324, 326, 327,
330, 331, 333, 334, 335
  parallel, 274, 276, 277,
    278, 283, 284, 297
  sequential, 274, 275, 281,
    282
  simultaneous, 275, 276,
    277, 283, 297
  software faults, xx, 111,
    120, 152, 158, 159, 285
Parallel DB Corp., 347
Partitioning, 39
Patterson, D., 121, 349
PdbReplicate, 347
Peer Direct Corporation, 348
Peer Direct Enterprise, 348
Pervasive Software, 348
Ping-ponging, 38, 64, 65, 135,
  225, 352
  prevention, 65
    partitioning, 67
Primary, 29, 338
Process pairs, 11, 12, 13, 16,
  43, 61, 115
  processor pairing, 14, 21
  random distribution, 16, 19,
    21
Quest Software, 348
Queuing, 52, 53
RAID (redundant arrays of
  independent disks), 168,
  169, 171, 228, 349
RDF, 126, 347, 351
Recovery, xx, xxiii, xxvi, 15,
  23, 41, 49, 60, 61, 74, 97,

103, 104, 105, 107, 112,
113, 116, 117, 118, 119,
120, 121, 122, 123, 124,
125, 126, 128, 129, 130,
133, 134, 137, 139, 140,
141, 143, 144, 145, 146,
149, 150, 151, 152, 155,
156, 158, 175, 177, 218,
221, 222, 249, 271, 272,
274, 275, 276, 277, 279,
280, 281, 282, 283, 284,
285, 305, 306, 326, 330,
333, 338, 346
  subsystem repair, 23, 116,
    117, 120, 273, 274, 277,
    281, 282, 284, 305
  system, 107, 116, 117, 118,
    120, 122, 123, 124, 134,
    272, 276, 277, 279, 283,
    284, 285, 305, 326, 330,
    333
Recovery point objective
  (RPO), xx, 124, 125, 126,
  127, 128, 129, 130, 133,
  139, 141, 143, 146, 147,
  149, 150, 151, 152, 178,
  221, 338, 351
Recovery time objective
  (RTO), xx, 124, 125, 126,
  127, 128, 129, 130, 133,
  137, 139, 146, 147, 149,
  150, 151, 152, 175, 178,
  221, 338, 351
redundant, 8, 9, 10, 22, 109,
  168, 177, 349
  triple redundancy, 167, 169
Referential integrity, xxii, 63,
  147, 178, 226, 229, 234,

236, 237, 238, 242, 245, 248, 251, 260, 264
foreign keys, 234, 236, 237
Reliability, 3, 5, 15, 22, 34, 40, 110, 349, 350, 351
   availability, xvii, xviii, xx, xxi, xxiii, xxiv, xxvi, xxvii, 4, 5, 6, 7, 8, 9, 10, 11, 12, 13, 14, 15, 16, 17, 18, 19, 21, 22, 23, 24, 25, 26, 27, 28, 29, 30, 31, 33, 34, 36, 40, 41, 42, 43, 44, 45, 46, 49, 60, 77, 78, 105, 107, 108, 109, 110, 114, 117, 118, 119, 120, 121, 122, 123, 124, 125, 150, 152, 153, 155, 156, 157, 158, 159, 161, 164, 165, 166, 167, 169, 172, 175, 176, 177, 178, 217, 220, 236, 271, 273, 279, 280, 281, 284, 285, 289, 293, 301, 303, 305, 309, 312, 315, 319, 322, 325, 328, 332, 335, 337, 342, 346, 350
   MTBF, xxiii, 4, 5, 7, 18, 19, 23, 36, 40, 41, 78, 108, 124, 156, 159, 161, 164, 167, 168, 176, 273, 293, 306
   MTR, xxiii, 4, 5, 7, 17, 18, 19, 23, 36, 40, 108, 110, 116, 122, 124, 156, 161, 176, 273, 274, 276, 293, 294, 306
   sparing, 15

system splitting, xviii, 27, 31, 43, 45, 77, 119, 150, 162, 295
Repair, xxiii, 4, 17, 18, 23, 36, 40, 105, 107, 108, 112, 113, 116, 117, 118, 120, 121, 122, 124, 125, 155, 156, 157, 164, 175, 177, 271, 272, 273, 274, 275, 276, 277, 278, 280, 281, 282, 283, 284, 285, 287, 288, 293, 294, 297, 305, 306, 309, 312, 315, 318, 321, 324, 326, 327, 330, 331, 333, 334, 335
Restoration, xx, xxiii, xxvi, 15, 41, 49, 60, 61, 97, 103, 105, 107, 112, 113, 116, 117, 118, 119, 120, 121, 122, 123, 124, 125, 126, 128, 129, 130, 133, 134, 137, 139, 140, 141, 143, 144, 145, 146, 149, 150, 151, 152, 155, 156, 158, 175, 177, 218, 221, 222, 249, 271, 272, 275, 276, 277, 279, 280, 282, 283, 284, 285, 305, 306, 326, 330, 333, 338, 346
Scalability, xxiv, 10, 93, 162
Shadowbase, xxvi, 347
SharePlex, 348
Sombers, xxv
Sparing, 15, 17, 20
Standby
   cold, 133, 139, 149, 218, 221
   hot, 133, 140, 144, 218, 222
   warm, 133, 140, 218

Standish Group, 24, 49, 111, 352
Storage Replicator, 348
Strickler, G., xxv, 38, 65, 135, 225, 351, 352
Sybase Replication Server, 348
Sybase, Inc., 348
Synchronous, xx, xxii, xxiii, 46, 54, 62, 68, 76, 77, 81, 82, 86, 89, 93, 97, 98, 102, 104, 136, 137, 141, 142, 143, 144, 145, 146, 147, 149, 152, 162, 163, 165, 169, 170, 171, 176, 177, 178, 182, 183, 185, 187, 188, 190, 200, 201, 202, 203, 205, 207, 208, 223, 224, 225, 226, 228, 230, 243, 293, 294, 298, 342, 347
System splitting, xviii, xix, xxiii, 26, 27, 31, 33, 34, 36, 37, 39, 41, 42, 43, 44, 45, 77, 78, 109, 119, 120, 121, 124, 150, 155, 157, 160, 162, 165, 175, 177, 283, 284, 295
Target database updating, 53
Transaction count limit
   partial transactions, 262
Transaction manager, 58, 64, 82, 83, 87, 177, 182, 194, 227, 230, 257
   two-phase commit protocol, 82, 83, 87, 143
Two-phase commit protocol, 82, 83, 87, 143
UNIX, xxiv, xxvi, 3, 8, 21, 22
Veritas Software, 348
Volume Replicator, 348
Wood, Alan, xxv, 4, 271, 352

About the Authors

Dr. Bill Highleyman, Paul J. Holenstein, and Dr. Bruce Holenstein have a combined experience of over 80 years in the implementation of fault-tolerant, highly available computing systems. This experience ranges from the early days of custom redundant systems to today's fault-tolerant offerings from HP (NonStop) and Stratus.

Dr. Bill Highleyman has done extensive work on the effect of failure mode reduction on system availability. He has built fault-tolerant systems for train control, racetrack wagering, securities trading, message communication, and other applications.

Paul J. Holenstein and Dr. Bruce Holenstein have architected and implemented the various data replication techniques required for the availability enhancements described in this book. Their company, Gravic, provides the ITI Shadowbase line of data replication products to the fault-tolerant community.